Simo '04

2003 EDITION

Statistics on Science and Technology in Europe

Data 1991-2002

9

EUROPEAN COMMISSION

Community Research

eurostat

.... Immediate access to harmonised statistical data

Eurostat Data Shops

A personalised data retrieval service

In order to provide the greatest possible number of people with access to high-quality statistical information, Eurostat has developed an extensive network of Data Shops ([1]).

Data Shops provide a wide range of **tailor-made services:**

- ★ immediate information searches undertaken by a team of experts in European statistics,
- ★ rapid and personalised response that takes account of the specified search requirements and intended use,
- ★ a choice of data carrier depending on the type of information required.

Information can be requested by phone, mail, fax or e-mail.

([1]) See the list of Eurostat Data Shops at the end of this publication.

Internet

Essentials on Community statistical news

- ★ Euro indicators: more than 100 indicators on the euro zone; harmonised, comparable, and free of charge.
- ★ About Eurostat: what it does and how it works.
- ★ Products and databases: a detailed description of what Eurostat has to offer.
- ★ Indicators on the European Union: convergence criteria; euro yield curve and further main indicators on the European Union at your disposal.
- ★ Press releases: direct access to all Eurostat press releases.

For further information, visit us on the Internet: www.europa.eu.int/comm/eurostat/

Europe Direct is a service to help you find answers to your questions about the European Union

New freephone number:

00 800 6 7 8 9 10 11

A great deal of additional information on the European Union is available on the Internet.
It can be accessed through the Europa server (http://europa.eu.int).

Cataloguing data can be found at the end of this publication.

Luxembourg: Office for Official Publications of the European Communities, 2004

ISBN 92-894-6823-8

Printed in France

PRINTED ON WHITE CHLORINE-FREE PAPER

The European Council in Lisbon in 2000 set the strategic goal of transforming the European Union by 2010 into ***the most competitive and dynamic knowledge-based economy in the world, capable of sustainable economic growth with more and better jobs and greater social cohesion***. Ever since, many decisions have been taken in order to implement the Lisbon Strategy. In particular, the ***Barcelona Council meeting in 2002*** set some clear and specific targets that would allow for monitoring progress and achievement of the more general goals. It was agreed that Member States should strive to achieve 3% of GDP to be spent on research by 2010 of which one third of this expenditure should be financed by the Government sector and two thirds by the Business sector. This strategy responds to the fact that ***Research and Development — R&D*** — is a key factor for boosting productivity, economic growth, employment and social cohesion. In this context, indicators are paramount to inform policy makers as to where Europe stands in terms of science and technology and how its position is evolving, also compared to Japan and the United States.

The statistics and indicators presented in this publication, jointly prepared by the ***Statistical Office of the European Communities — Eurostat — and the Research Directorate General of the European Commission***, report on Europe's recent performance, allowing a close monitoring of the situation and the identification of current and potential areas of concern. This edition of ***Statistics on Science and Technology***, as with the 2000 edition, marks a departure from the customary format allowing the presentation of recent developments not only in the four areas looked at each year — R&D expenditure, R&D personnel, Government budget appropriations or outlays on R&D and Patents, but also in more recent areas of interest to science and technology policy-makers and analysts. Thus, data on Human resources in science and technology and on high tech sectors are also presented in this publication.

Also, where data were available, gender main-streamed statistics were produced for most indicators. The accession of ten new Member States in 2004 will add an additional dimension to the strategic goal set at Lisbon; these countries will also have to increase their efforts to achieve the targets defined at the ***Barcelona Council meeting***. Each chapter, therefore, contains statistics on the ten Acceding and three Candidate Countries.

This edition of ***Statistics on Science and Technology in Europe*** has been jointly managed and prepared by two services of the European Commission: ***Eurostat***, Unit B-5, headed by ***Sylvie Ribaille***, and ***DG Research***, Unit K-3, headed by ***Ugur Muldur***. ***Statistics on Science and Technology in Europe*** complements regular publications of each service, such as Eurostat's ***Science and Technology in Europe — Statistical pocketbook*** and provides a wide ranging set of statistics which are comprehensive, internationally comparable and as up to date as possible and comments on them. It also complements the more policy-oriented analyses based on S&T indicators publications produced by the Research Directorate General such as the ***Third European Report on Science and Technology Indicators*** and ***Key Figures on the European Research Area***.

By addressing policy needs in this important area of ***Science and technology statistics***, regular publications such as the ***Panorama report*** fulfil a significant role. However, there still remains a lot of work to be done dealing with developing pertinent new indicators, harmonising and improving existing indicators and making international data internationally comparable and available. We hope that this report will be welcomed as a useful tool for the policy-making community and all who analyse ***Science and Technology — S&T***.

Sylvie Ribaille
Head of Unit B-5
Eurostat

Ugur Muldur
Head of Unit K-3
Directorate-General for Research

This publication was jointly prepared by two services of the European Commission:

- Eurostat, Unit B-5, Head of Unit: Sylvie Ribaille and
- the Directorate-General for Research, Unit K-3, Head of Unit: Ugur Muldur.

This edition of the *Panorama report* was coordinated and managed by Fabienne Corvers, Simona Frank, August Götzfried and Ian Perry, with the collaboration of Kai Husso, Dermot Lally, Brian Sloan, Guido Strack and David Uhlir.

The technical work was carried out by CAMIRE and financed under the study contract number ERBHPV2-CT-2001-90002 which formed part of the Common Basis of Science, Technology and Innovation Indicators – *CBSTII* – activity managed by Unit K-3, of the *5th Framework Programme for Research and Technological Development*.

The texts and analyses were realised by:

- Marta Alfageme Perez de Mendiguren, Amina Kafai, Sammy Sioen, Alex Stimpson and Christophe Zerr, under the responsibility of Elisabeth Lamp.

The data processing, the layout of the publication and the desktop publishing were undertaken by:

- Marie-Agnès Bragard, Gérard Carlier and Damien Tornaboni.

DISCLAIMER

The opinions expressed in this publication are those of the individual authors alone and do not necessarily reflect the position of the European Commission.

Contributions

Eurostat and DG Research gratefully acknowledge the contributions made by the following institutes that supplied the statistics on the different countries:

- **Belgium** .... Federal Office for Scientific, Technical and Cultural Affairs;
  .... Institut National de Statistique;
- **Denmark** .... The Danish Institute for Studies in Research and Research Policy;
  .... Danmarks Statistik;
- **Germany** .... Bundesministerium für Bildung, Wissenschaft, Forschung und Technologie – BMBF;
  .... Statistisches Bundesamt;
  .... Zentrum für Europäische Wirtschaftsforschung, GmbH – ZEW;
- **Greece** .... General Secretariat for Research and Technology;
- **Spain** .... Instituto Nacional de Estadística – INE;
- **France** .... Ministère de l'Éducation Nationale, de la Recherche et de la Technologie – MENRT;
  .... Ministère de l'Économie, des Finances et de l'Industrie;
- **Ireland** .... Forfas;
- **Italy** .... Istituto Nazionale di Statistica – ISTAT;
  .... Consiglio Nazionale delle Ricerche – CNR;
  .... Instituto di ricerca sull'empresa e lo sviluppo – CERIS;

- **Luxembourg** . . . . Service Central de la Statistique et des Études Économiques – STATEC;
  . . . . Centre d'Études de Populations, de Pauvreté et de Politiques Socio-Économiques/
  . . . . International Networks for Studies in Technology, Environment, Alternatives,
  . . . . Development – CEPS/INSTEAD;
- **Netherlands** . . . . Statistics Netherlands – CBS;
- **Austria** . . . . Statistics Austria;
- **Portugal** . . . . Observatório das Ciências e das Tecnologia – OCT;
- **Finland** . . . . Statistics Finland;
- **Sweden** . . . . Statistics Sweden;
- **United Kingdom** . . . . . Office for National Statistics – ONS;
  . . . . The Department of Trade and Industry – DTI;
- **Czech Republic** . . . . . Czech Statistical Office;
- **Estonia** . . . . State Statistical Office of Estonia;
- **Cyprus** . . . . Statistical Service of Cyprus;
- **Latvia** . . . . Central Statistical Bureau of Latvia;
- **Lithuania** . . . . Statistics Lithuania;
- **Hungary** . . . . Hungarian Central Statistical Office;
- **Malta** . . . . National Statistics Office;
- **Poland** . . . . Central Statistical Office of Poland;
- **Slovenia** . . . . Statistical Office of the Republic of Slovenia;
- **Slovak Republic** . . . . . Statistical Office of the Slovak Republic;
- **Bulgaria** . . . . National Statistical Institute;
- **Romania** . . . . National Institute of Statistics;
- **Turkey** . . . . State Institute of Statistics;
- **Iceland** . . . . The Statistical Bureau of Iceland;
- **Norway** . . . . Statistics Norway;
  . . . . Norwegian Institute for Studies in Research and Higher Education – NIFU.

In addition, Eurostat and DG Research would like to thank the following institutions:

- **European Patent Office – EPO;**
- **Organisation for Economic Co-operation and Development – OECD.**

Maps

GISCO, Eurostat.

List of tables, figures and maps

Maps

Figures

Maps

Maps

This publication presents an analysis of ***Science and Technology in Europe*** by looking at the main statistical indicators in this field. The publication, intended for both generalists and specialists, is organised in four parts:

- The first part presents an analysis of key R&D input indicators: Government Budget Appropriations or Outlays on R&D — GBAORD — and R&D expenditure.
- In Part 2, an insight into the human resources in R&D and the attractiveness of S&T professions in Europe is given.
- Part 3 presents S&T performance indicators, as it looks at statistics on patents and the development of high tech sectors.
- Finally, the last part provides accompanying methodological information in some detail for more specialist users.

The primary focus of the statistics and indicators presented in this publication is on the 15 European Union Member States and, to a lesser extent, the European Economic Area — EEA. On the eve of the accession of 10 new countries into the EU, this publication also looks at the Acceding and Candidate Countries, whenever data are available and reliable. To provide high-level international comparison, data for the United States and Japan are also presented, where possible. At the other end of the scale, a regional analysis within the EU countries is provided. Data are as comprehensive, comparable and as up to date as possible.

Given the numerous sources of data involved, the coverage of the time series differs according to indicator. However, the first year taken into consideration for most indicators in this publication is 1991. In any case, the goal of this publication remains the same throughout: to provide the most detailed and coherent time series analysis possible.

Consistency with the reporting presented in previous publications is also maintained, whilst seeking to complement these aspects with further research that responds to user requirements.

Due to constraints on space, the comprehensive statistical tables used for the analyses are not always available in the paper version of the present publication. Instead, they may be seen in their entirety in:

- the CD-ROM ***Statistics on Science and Technology 2003*** or in
- Eurostat's reference database ***NewCronos***.
- The ***Third European Report on Science and Technology Indicators — 2003*** can be integrally downloaded at http://www.cordis.lu/indicators.

Government budget appropriations or outlays on R&D

Chapter 1 shows that in 2001, budget appropriations for R&D totalled roughly EUR 69 thousand million in the Member States of the European Union, EUR 30 thousand million in Japan and EUR 104 thousand million in the United States. As a percentage of GDP, GBAORD in EU-15, Japan and the United States amounted to 0.75%, 0.71% and 0.94%, respectively. Although GBAORD in the EU and the United States decreased in the early 1990s, annual average real growth rates for the 1997-2002 period reveal an increasing trend in all the EU, Japan and the United States.

Within the EU, France and Finland show the highest proportions of government budgeting to R&D activities, both as a proportion of their respective GDP and total general government expenditure. Luxembourg, Spain and Ireland, on the other hand, have shown the highest annual average real growth rates during the 1997-2002 period, but their GBAORD as a percentage of GDP figures are still below the EU average.

If the distribution of GBAORD by socio-economic objective is taken into account, 'Research financed from General University Funds (GUF)' accounted for the lion's share of EU-15's GBAORD, as it represented 32.4% of the total. 'Research financed from GUF' was also the main socio-economic objective in Japan with 34.9% of total appropriations. However, in the United States over half of total GBAORD in 2002 was allocated to 'Defence' (54.0%). Among Member States, the distribution by socio-economic objective varies: 'Defence' is the main or one of the main objectives for Spain (30.2%), France (24.2%), Sweden (22.2%) and the United Kingdom (34.9%), but it represents less than 7% of national total GBAORD for the rest of the countries. For certain countries, such as Ireland, Iceland and Portugal, the objective 'Agricultural production and technology' is quite significant due to the importance of fishing activities in Iceland and agricultural activities in Ireland and Portugal. The 'Industrial production and technology' objective represented a noteworthy part of GBAORD in Belgium, Germany, Spain, Ireland, Italy, Portugal and Finland. In Iceland, the main socio-economic objective was 'Social structure and relationships'.

R&D expenditure

Chapter 2 gives the most recent trends in R&D expenditure. According to Eurostat estimations, the EU increased its R&D intensity to 1.99% in 2002, the gap with respect to Japan (2.98% in 2000) and the United States (2.80%) remaining considerable. Looking at the estimations by institutional sector, in 2002, R&D expenditure as a percentage of GDP in EU-15 amounted to 1.30% in the Business enterprise sector – BES, 0.42% in the Higher education sector – HES – and 0.26% in the Government sector – GOV. The trends were slightly up for the GOV and HES but remained stable for the BES.

According to data for the latest year available for each country, the top three Member States of the EEA in terms of R&D intensity were Sweden (4.27% in 2001), Finland (3.49% in 2002) and Iceland (3.11% in 2002). For Acceding Countries overall, a rate of R&D expenditure as a percentage of GDP equal to 0.84% was estimated in 2001. With figures above 1.30%, the Czech Republic and Slovenia were leading, other countries retaining percentages below 1%.

In 2002, the EU spent 151 thousand million of constant 1995 PPS (EUR 182 thousand million) on R&D, recording an annual real growth rate of 1.90% compared to the previous year. At the national level and according to the latest available data for each country, the top three countries in R&D expenditure in absolute terms were Germany with EUR 52 thousand million, France (EUR 33 thousand million) and the United Kingdom (EUR 30 thousand million). The highest real growth rates in R&D expenditure were registered by Portugal (8.6%), Denmark (7.4%) and Ireland (7.4%). More than half of the R&D expenditure recorded by the Acceding Countries was carried out in Poland and the Czech Republic, as in 2001 they spent EUR 1 323 million and EUR 832 million respectively on R&D.

At the regional level, in 2001 the top 10 regions in the EU in terms of R&D expenditure as a percentage of GDP were mainly located in Germany, Sweden and Finland. The German region Braunschweig leads with 6.21% of the GDP devoted to R&D followed by Västsverige (SE) and Stuttgart (DE) with 5.27% and 4.82%, respectively.

In 2001, 30% of the EEA's R&D expenditure was concentrated in ten regions when measured in constant 1995 PPS, located mainly in Germany, France, Denmark, Italy and Sweden. Most R&D expenditure was carried out in Île de France (FR), where it accounted for 8.1% of the EEA total. Following Île de France were Oberbayern (DE) with 4.1% and Stuttgart (DE) with 3.5%.

R&D personnel

As documented in Chapter 3, in 2002, 1.39% of the labour force in EU-15 was employed in R&D. At the national level, Iceland leads among EEA countries with 3.09% of its labour force employed in R&D, ahead of Finland 2.60%, Sweden 2.43% and Denmark 2.11%. Expressed in full-time equivalent – FTE, Eurostat estimated that 1.83 million people worked in R&D within the EU in 2002, which represented an increase of 1.6% compared to 2001. The breakdown by institutional sector showed that 55% of the R&D personnel in EU-15 were employed in the BES, 30% in the HES and 14% in the GOV.

The share of R&D personnel in the labour force was 0.84% in the Acceding Countries. The two Acceding Countries with the highest figures were Slovenia and Hungary, with 1.36% and 1.11% of their labour force employed in R&D, respectively.

According to the latest available data on researchers measured in FTE, the United States employed the highest number of researchers (1 114 100 people in 1997), compared to the EU (1 001 209 in 2002) and Japan (647 572 in 2000). Women researchers are less well represented than men, particularly in the BES. In 2001, they accounted for 44% of total researchers in Portugal and 39% in Iceland, where the maximum values were obtained among EEA countries.

In general, the proportion of researchers by field of science is higher in the HES than in the GOV for any given country or field of science. The highest researcher proportions are registered in the fields of 'Medical sciences' and 'Social sciences'.

In the EEA, one quarter of the R&D personnel in FTE is concentrated in 10 European regions. Accounting for 7.0% of the EEA's total R&D personnel, Île de France (FR) was the leader, followed by Oberbayern (DE) and Stuttgart (DE), with 3.5% and 2.6%, respectively.

Human resources in science and technology

Chapter 4, on Human Resources in Science and Technology, shows that the number of students taking tertiary education courses is growing, both in the EU and the Acceding Countries. In 2001, over ten million people in the EU were following tertiary education courses, as were just over 2.9 million in the Acceding Countries. Science and engineering courses together accounted for just over a quarter of all tertiary studies in the EU. Though women represented more than half of all students in practically every country, engineering courses, and to a lesser extent science courses, seem to have problems attracting women. Whilst women account for 53% of the EU's total participation in tertiary education, they only represented 22% in engineering courses and 38% in science courses.

The output of Europe's tertiary education institutions was close to 2 million new graduates in 2001 in the EU and around 631 thousand in the Acceding Countries. This compared with just over 1 million new graduates in Japan and over 2.1 million in the United States. In relative terms, with 55 graduates per thousand population aged 20-29, the Acceding Countries perform better, on average, than the EU (40). Women accounted for an even higher proportion of all graduates than they did for students. On average, 55.9% of all graduates were women in the EU in 2001, compared with 63.7% in the Acceding Countries, 49.4% in Japan and 57.0% in the United States.

The proportion of the EU's working population in S&T occupations in 2002 was around 28.6%, whereas it amounted to 25.4% in the Acceding Countries.

One fifth of all EU-15 25-64 year olds have tertiary education, compared to 15% in the Acceding Countries, but the age distribution of S&T workers is more skewed towards the younger population in the Acceding

Countries. Being a scientist or engineer was, on average, less common in the Acceding Countries than in the EU. But whilst this may be true, the gender balance between those people in the labour force working in S&E was far better.

As regards to HRST intensity by sector of activity, services have far more S&T workers than manufacturing. Other knowledge-intensive services, which includes 'Education' and 'Health and social work', has the highest ratio of tertiary educated employed people in both the EU and the Acceding Countries.

At the regional level and ranked according to the percentage of people in the labour force who are HRST, Stockholm (SE) is the leading region in the EU, where compared to the labour force over half of all residents either had a third level education or worked in S&T in 2002 (53.1%).

Patents

Chapter 5 demonstrates how patent applications to the ***European Patent Office*** – EPO – and patents granted by the ***United States Patent and Trademark office*** – USPTO – have been increasing during the nineties. The EPO received 60 890 patent applications from EU Member States, 168.3% of its value in 1996 and more than double the applications made in 1991. Patent applications to the EPO from Japan and the United States in 2001 amounted to 22 226 and 47 202 respectively. These represented 175.8 and 167.8% of their corresponding values in 1996. When taking population into account, the differences across the three blocks become smaller and the positions are inverted. In 2001, the highest ratio was registered by Japan – 175 patent applications per million inhabitants, followed by the United States (170) and the EU (161).

The USPTO granted 89 636 patents to US inventors in 2001, 30 285 to inventors from the EU and 33 733 to inventors from Japan. As a proportion of population, differences still remain large at the USPTO with 322 patents granted per million inhabitants for the United States, 265 for Japan and 80 for the European Union. The main explanations for these divergences are supposed to originate from the home advantage phenomenon which for the EU is, due to the additional national level, smaller than for the United States.

In order to overcome comparability problems associated to data derived from patents filed at a single patent office, the concept of patent family has been developed. In 1998, the patentees from the United States registered the highest number of triadic patent families (14 255), closely followed by the EU (13 187) and Japan (10 033).

Within the EU and in absolute terms, Germany is leading, accounting for 41.9% of total EPO applications in 2001, followed by France (14.1%) and the United Kingdom (13.1%). In relative terms, the country with the highest number of patent applications per million inhabitants was Sweden (367) followed by Finland (338). With rates that were twice as high as those for the EU average and the United States, both Sweden and Finland outperformed Germany, France and the United Kingdom in relative terms.

Among the patent applications to the EPO, an increasing proportion relates to high technology areas. Throughout the period 1996-2001, high tech patent applications in EU-15 grew at an annual average growth rate of 22.3%, compared to 11.0% of patent applications overall. This increase for high tech patents was evident not just for the EU, but also for patent applications made to the EPO by Japan and the United States. With 136 high tech patent applications per million inhabitants, Finland leads in the EU.

At the regional level, inventors from the French capital region of Île de France applied for most patents to the EPO in absolute terms (3 423 patent applications), whereas Oberbayern (DE) was the region with the highest proportion of EPO applications per million inhabitants (824) in the EU. In the high tech fields, Oberbayern led in absolute terms (1 138 patent applications) and Noord-Brabant (NL) as a proportion of population (342 patent applications per million inhabitants).

Europe's high tech sectors
Overview in terms of employment and trade

Chapter 6 provides an insight into how the EU performs in the high technology sectors, by looking at their contribution to employment, value added and external trade.

Employment in high tech sectors

With 163 million people employed in the EU, services sectors accounted for 68% of total employment in 2002. Among services, ***knowledge-intensive services*** – KIS – are becoming increasingly important (33.3% of total employment). Whilst high tech and medium-high tech manufacturing sectors account for 7.4% of employment, other manufacturing sectors employ 11.8% of the EU's workforce and other sectors (neither manufacturing nor services, i.e. agriculture, fishing, mining, construction, etc.) 12.9%.

At the Member State level, Germany was the country where high tech and medium-high tech manufacturing sectors accounted for the largest proportion (11.4%) of the national total employment. As for knowledge-intensive services, Sweden was most specialised in these sectors (47.0% of total employment).

At the regional level, the top 15 regions represented 31% of the EU's total employment in high tech and medium-high tech manufacturing sectors. Whilst Lombardia (IT) employed the highest number of people in high tech and medium-high tech sectors among EU regions (431 thousand), Stuttgart (DE) was the region most specialised in high tech and medium-high tech manufacturing sectors (21.2% of employment).

In knowledge-intensive services – KIS, the leading EU region in absolute terms in 2002 was Île de France (FR), as it employed 2 352 thousand people in these sectors. As a proportion of total employment, Inner London (UK) was ahead (59.1% of employment in KIS).

Value added and labour productivity

In terms of labour productivity, whilst that for overall manufacturing in the EU was EUR 52 thousand per person employed, high tech manufacturing sectors registered a rate of EUR 73 thousand, high tech services 68 thousand per person employed and medium-high tech manufacturing sectors EUR 58 thousand. Knowledge-intensive market services retained a labour productivity rate of 53 thousand per person employed, which was below the manufacturing average.

High tech trade

In 2001, trade of high technology products contributed to around a fifth of total trade in the EU with EUR 195.5 thousand million of high tech exports and EUR 218.6 thousand million of high tech imports accounting for 19.8% and 21.3% of total exports and imports, respectively. In comparison, the corresponding export proportions are higher in the United States and Japan (28.6% and 24.7%, respectively) whereas the import proportions are slightly lower, being 18.5% for both the United States and Japan. In the Acceding Countries, the high tech proportion of total exports was 9.7% while high tech imports accounted for 13.8% of total imports.

In 2001 the United States led in high tech trade closely followed by the EU; however, the EU had the largest high tech trade deficit of EUR 23 thousand million (excluding intra-EU trade). Although Japan was the third leading high tech exporter, it had the highest trade balance surplus of EUR 39 thousand million. Between 1996 and 2001, the EU high tech exports grew at an annual average growth rate of 15.0% slightly higher than in the United States (12.6%) and Japan (6.7%). The Acceding Countries experienced the highest growth in exports, reaching a rate of 33.9%.

In terms of world market share of high tech exports, again the United States remains ahead holding 18.0% of the market followed by EU with 15.0% and Japan 8.5%. The top six countries together represented 57.4% of the world's high tech exports market share. Among the EU countries, Germany, France, the United Kingdom and the Netherlands feature in the top exporters and importers of high tech products, all with a positive high tech trade balance when both intra and extra-EU trade are considered.

In 2001, intra-EU high tech exports accounted for nearly 60% of total EU high tech exports. Excluding intra-EU flows, a total of EUR 195.5 thousand million of high tech products were exported from the EU in 2001. 28.8% of these products were exported to the United States and 4.2% to Japan. With regard to imports, just over half of the total EU high tech imports originated from within the Member States. Among the EU Member States, the countries with the highest high tech trade deficits when only extra-EU trade is taken into account, were the Netherlands, the United Kingdom and Germany. France, Sweden and Finland had the highest high tech trade surpluses. In terms of exports of high tech trade products within the EU in 2001. The three main export partners of the Member States were Germany, the United Kingdom and France.

Between 1996 and 2001, high tech exports from the Acceding Countries quadrupled in value from EUR 3.4 thousand million to EUR 14.5 thousand million, registering an annual average growth rate of 33.9% during the 1996-2001 period. High tech imports, in turn, tripled from EUR 9.4 thousand million to EUR 25.8 thousand million. The EU was by far the largest high tech trade partner for Acceding Countries.

The distribution of high tech trade by product group in 2001 shows that 'Electronics' are by far the most traded goods in the EU, followed by 'Computer & office machinery' and 'Aerospace' products.

PART 1

Investing in R&D in Europe

1.1. Introduction

Statistics on Government budget appropriations or outlays on R&D – GBAORD – provide an idea on how governments support R&D activities. GBAORD includes all appropriations allocated to R&D in central or federal government budgets. Provincial or state government should be included only where the contribution is significant. Unless otherwise stated, the data include both current and capital expenditure, and cover not only government-financed R&D performed in government establishments, but also government-financed R&D in the business enterprise, private non-profit and higher education sectors, as well as abroad, e.g. in international organisations. Data are collected according to the guidelines outlined in the OECD's *Proposed standard practice for surveys of research and experimental development – Frascati Manual*, 2002.

GBAORD data do not take into account the amount of money actually spent, but are rather based on budget provisions, and so should be seen as intentions to spend. This is why data on actual R&D expenditure – see Chapter 2 – which are not available in their final form until some time after the end of the budget year concerned, may well differ from the original budget provisions. The process of political consensus on public expenditure creates gaps between budgets and final expenditure, both in terms of time and amount of resources. The reporting unit also differs between GBAORD and R&D expenditure: whilst the reporting unit for GBAORD is the Government, that for R&D expenditure is the performer of the R&D activity. However, since there is a greater time lag for obtaining final R&D expenditure data, these are usually also collected from budget statistics in order to provide timely indicators.

Data are collected at the national level and the procedure is articulated in a two step process:

- within the budget statistics, it is first necessary to identify the budget items that involve R&D;
- the R&D content of these budget items must then be measured or estimated.

GBAORD data reflect policies at a given moment in time and the concomitant priorities of the policy makers when allocating their budgets. These data are hard to collect because they are not obtained from *ad-hoc* surveys, but in most cases are obtained from national budget statistics. The difficulty is due more specifically to the fact that national budgets already have their own terminology and methodology and therefore do not accord entirely with the Eurostat guidelines and the methodology proposed by the *Frascati Manual*.

Government R&D appropriations are broken down by socio-economic objectives on the basis of the NABS classification – *Nomenclature for the analysis and comparison of scientific programmes and budgets*.

The chapter is divided into two main sections:

- Section 1.2. takes an international perspective and compares the present status and development of GBAORD in the EU, Japan and the United States.
- Section 1.3. focuses on the evolution of GBAORD in the Member States of the EU, Candidate Countries, Iceland and Norway.

The analysis in this chapter covers the period 1992 to 2003, with 2003 data being provisional. Readers should notice that, as no data were available for Luxembourg until 2000, EU-15 totals in this chapter include Luxembourg only from 2000 onwards. For further information on the methodology used, please refer to methodological notes starting on page 150.

Figure 1.1. **GBAORD in constant 1995 PPS EU-15, Japan and the United States 1992 to 2002 (1,2)**

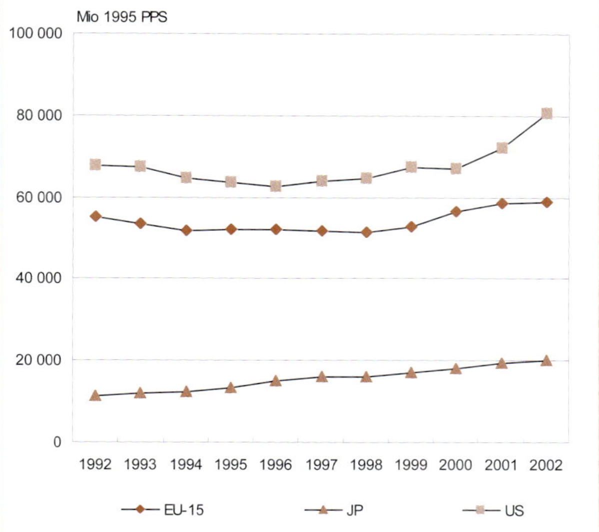

(1) EU-15 1995-96 and 2000-2002: Eurostat estimates.
(2) JP and US 2002: provisional data.

Sources: Eurostat; OECD.

Figure 1.2. **GBAORD as a % of GDP EU-15, Japan and the United States 1992 to 2002 (1,2)**

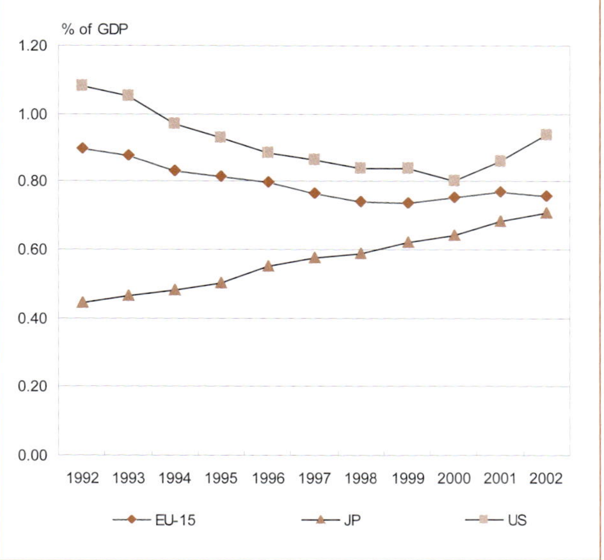

(1) EU-15 1995-96 and 2000-2002: Eurostat estimates.
(2) JP and US 2002: provisional data.

Sources: Eurostat; OECD.

1.2. An international perspective: EU-15, Japan and the United States

This section considers government budgeting for R&D activities in the European Union compared to that of Japan and the United States. Overall levels of GBAORD are examined as well as its breakdown by socio-economic objective.

Total GBAORD

In 2002, the largest amount of funds for R&D activities was allocated by the US Government, followed by EU-15 and Japan

The importance attributed to R&D by the three major economies has changed over the last ten years – Figure 1.1. In 2002 total GBAORD amounted to approximately 59 thousand million of 1995 constant PPS in the EU, 20 thousand million in Japan and exceeded 80 thousand million in the United States. In terms of nominal value (current EUR), GBAORD amounted to approximately EUR 104 thousand million, EUR 69 thousand million and EUR 30 thousand million in the United States, EU-15 and Japan respectively.

However, as shown in Figure 1.2., differences are less important when GBAORD is considered as a percentage of GDP. In 2002, GBAORD in EU-15, Japan and the United States amounted to 0.75%, 0.71% and 0.94% of their GDP, respectively. In 1993, GBAORD as a percentage of GDP for the EU was 1.9 times larger than that of Japan, whereas that of the United States was 2.3 times higher. During the late 1990's, there was a GBAORD convergence at the international level, the EU, Japan and the United States reaching a similar level in 2000 – 0.75%, 0.64% and 0.80% of their respective GDP.

Between 1992 and 2002, Japan's GBAORD rose by 78% in real terms (constant 1995 PPS) whereas EU-15's remained quite stable. In the United States, GBAORD in real terms remained fairly stable between 1992 and 2000 but increased by 20% between 2000 and 2002.

Between 1992 and 1997, GBAORD decreased in the European Union and in the United States with annual average growth rates of -1.3% and -1.2%, respectively – calculated in constant 1995 PPS. On the contrary, GBAORD in Japan increased at an annual average growth rate of 7.1% during the 1992-97 period – Table 1.1.

After 1997, the trend of GBAORD in the EU and the US changed, and between 1997 and 2002, it increased at an annual average growth rate of 2.7% in the European Union and 4.7% in the United States and Japan.

GBAORD by socio-economic objective

'Research financed from General University Funds (GUF)' accounts for the lion's share in EU-15 and Japan whereas 'Defence' does so in the United States

On the basis of the NABS classification, GBAORD is broken down by socio-economic objective, corresponding to the specific aims of the appropriations or outlays.

Not only does the level of budgeting for R&D activities vary from one geographical entity to another, but the objectives are also different. Figure 1.3. displays these different approaches to budgetary appropriations for the EU, Japan and the United States. It may be seen that in 2002, as in previous years, 'Research financed from General University Funds (GUF)' accounted for the lion's share of EU-15's GBAORD as it amounted to 32.4% of total GBAORD.

In Japan, 'Research financed from GUF' was also the main socio-economic objective, in 2002, with 34.9% of total appropriations. Two other objectives – 'Production, distribution and rational utilisation of energy' and 'Non-oriented research' – also accounted for more than 15% of Japan's total GBAORD.

In the United States, over half of total GBAORD in 2002 was allocated to 'Defence' (54.0%). The second main objective was 'Protection and improvement of human health', representing 24.9% of the total GBAORD.

Looking at the annual average growth rates by socio-economic objective – Table 1.1., it may be seen that civil appropriations increased faster than total appropriations between 1997 and 2002 in the EU, Japan and the United States. This means that the objective 'Defence' increased at a lower rate than total appropriations – in the EU and the United States – and decreased – in Japan.

In the European Union, the objectives that increased the most between 1997 and 2002 are 'Social structures and relationships' and 'Other civil research', as they recorded annual average real growth rates of 9.0% and 8.1%, respectively.

In Japan, four objectives had annual average growth rates above 10%. These were 'Exploration and exploitation of the earth', 'Infrastructures and general planning of land use', 'Control and care of the environment' and 'Non-oriented research'.

In the United States, the objectives that increased the most were 'Non-oriented research' and 'Protection and improvement of human health'. On the contrary, 'Production, distribution and rational utilisation of energy' decreased by almost 10%.

Government Budget Appropriations or Outlays on R&D — GBAORD

Table 1.1. **Annual average real growth rates of GBAORD by socio-economic objective in % EU-15, Japan and the United States 1992 to 1997 and 1997 to 2002 (1,2)**

	Socio-economic objectives	EU-15		JP		US	
		1992-97	1997-2002	1992-97	1997-2002	1992-97	1997-2002
1.	Exploration and exploitation of the earth	-7.2	2.7	12.3	10.6	-2.4	4.4
2.	Infrastructure and general planning of land use	-2.2	2.6	15.2	14.2	0.0	-2.6
3.	Control and care of the environment	0.9	3.3	8.6	13.9	1.1	-0.9
4.	Protection and improvement of human health	4.5	3.5	14.6	4.2	2.5	11.7
5.	Production, distribution and rational utilisation of energy	-2.8	-0.6	5.9	1.6	-7.2	-9.7
6.	Agricultural production and technology	-1.4	-1.2	5.8	5.4	0.0	3.3
7.	Industrial production and technology	-6.5	4.5	19.1	7.5	13.9	0.0
8.	Social structures and relationships	-2.1	9.0	5.2	1.6	-7.0	3.7
9.	Exploration and exploitation of space	-1.4	-0.4	4.6	3.7	0.9	-5.1
10.	Research financed from General University Funds (GUF)	1.1	3.4	4.4	3.3	:	:
11.	Non-oriented research	2.3	2.3	13.0	12.4	0.0	12.3
12.	Other civil research	-8.8	8.1	:	:	:	:
13.	Defence	-4.9	0.9	6.7	-2.6	-2.3	4.3
	Total civil appropriations	-0.5	3.0	7.2	5.1	0.4	5.3
	Total appropriations	**-1.3**	**2.7**	**7.1**	**4.7**	**-1.2**	**4.7**

NB: Annual average growth rates — AAGR — are calculated in constant 1995 PPS.
(1) EU-15 2002: Eurostat estimates.
(2) JP and US 2002: provisional data.

Sources: Eurostat; OECD.

Figure 1.3. **Distribution of GBAORD by socio-economic objective in % EU-15, Japan and the United States 2002 (1,2)**

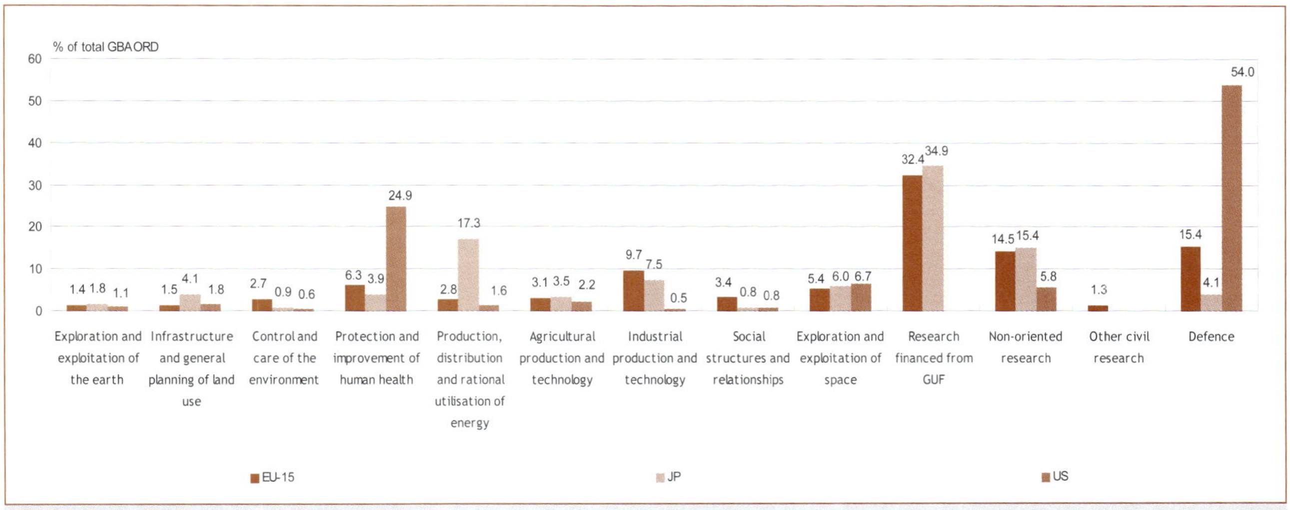

NB: JP — Total excludes other civil research;
US — Total excludes GUF and other civil research.
(1) EU-15 2002: Eurostat estimates.
(2) JP and US 2002: provisional data.

Sources: Eurostat; OECD.

1.3. A European perspective: EU-15, Candidate Countries, Iceland and Norway

Total GBAORD

Iceland (1.14%) and France (1.03%) allocate the highest percentage of their GDP to R&D activities

According to estimations made by Eurostat on the basis of provisional data, in 2002 governments of the European Union allocated around EUR 69 thousand million in budget appropriations or outlays for R&D. In real terms (constant 1995 PPS) this represented approximately 59 thousand million constant PPS.

As shown in Figure 1.4., in nominal terms (current EUR), GBAORD of the European Union decreased between 1992 and 1994 but increased between 1994 and 2002. In real terms, GBAORD of the European Union decreased until 1998 which marked the lowest point at 52 thousand million of constant 1995 PPS. From 1998 onwards, GBAORD showed a slight increase and reached 59 thousand million of constant 1995 PPS in 2002.

Whilst EU GBAORD represented 0.75% of its total GDP in 2002, this figure conceals differences between the Member States as demonstrated by Figure 1.5.

The greatest efforts in terms of R&D funding were made by Iceland, France and Finland, with 1.14%, 1.03% and 0.98% of their GDP allocated to R&D. In Sweden, Norway and Germany, there was also more emphasis placed on government budgeting for R&D activities than the EU-15 average of 0.75%.

Ireland, Greece and Luxembourg, on the other hand, made smaller budgetary appropriations for R&D: 0.33 %, 0.28% and 0.25% of their respective GDP, being around two-fifths of the Community average. The eight other EEA countries fell within a bracket between 0.73 % (The Netherlands) and 0.59% (Austria) of GDP.

Within the Candidates Countries, it is Slovenia that allocated the most to R&D activities with 0.53% of GDP. On the contrary, in Romania GBAORD only amounted to 0.16% of its GDP.

GBAORD as a percentage of total general expenditure is another indicator that provides an estimation of the relative emphasis that governments place on publicly funded R&D – Figure 1.6. The three countries – Iceland, France and Finland – that had the highest GBAORD as a percentage of GDP also had the highest GBAORD as a percentage of total general expenditure in 2002: 3.00% in Iceland, 1.92% in France and 2.02% in Finland. These countries were followed by Spain – 1.73% of total general expenditure, the Netherlands and the United Kingdom – both with 1.70%. At the other extreme, Luxembourg only allocated 0.46% and Greece 0.59%.

Figure 1.4. **GBAORD in current EUR and in constant 1995 PPS EU-15 1992 to 2002 (1)**

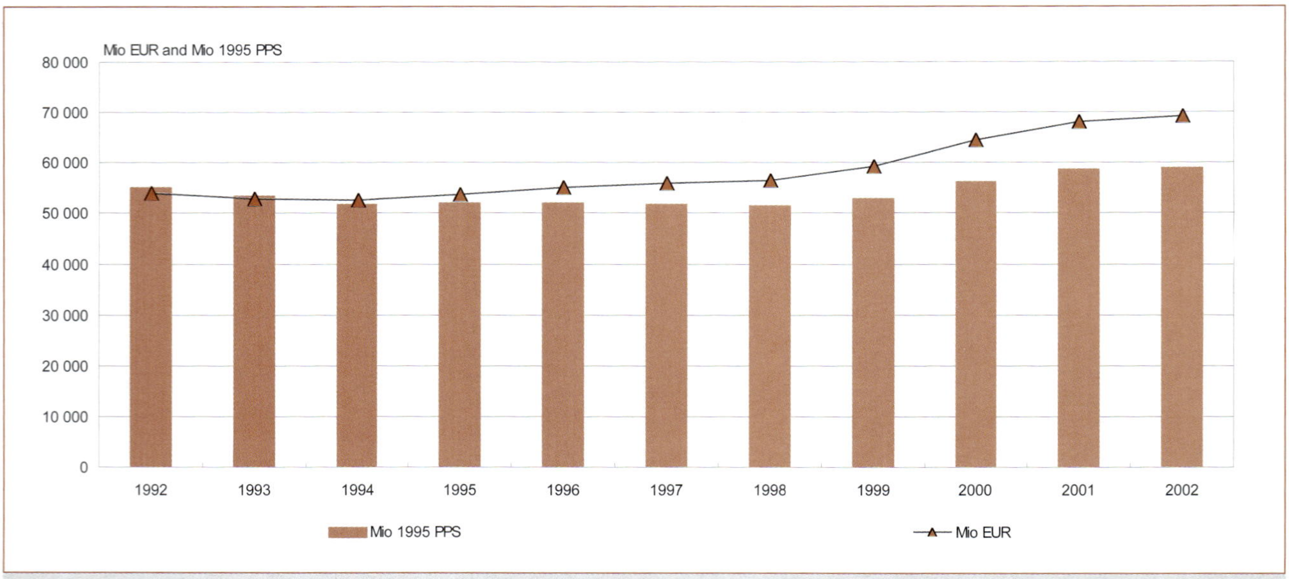

(1) EU-15 1995-96 and 2000-2002: Eurostat estimates.

Source: Eurostat.

Figure 1.5.

GBAORD as a % of GDP
EU-15, Candidate Countries, Iceland and Norway
2003 (1, 2)

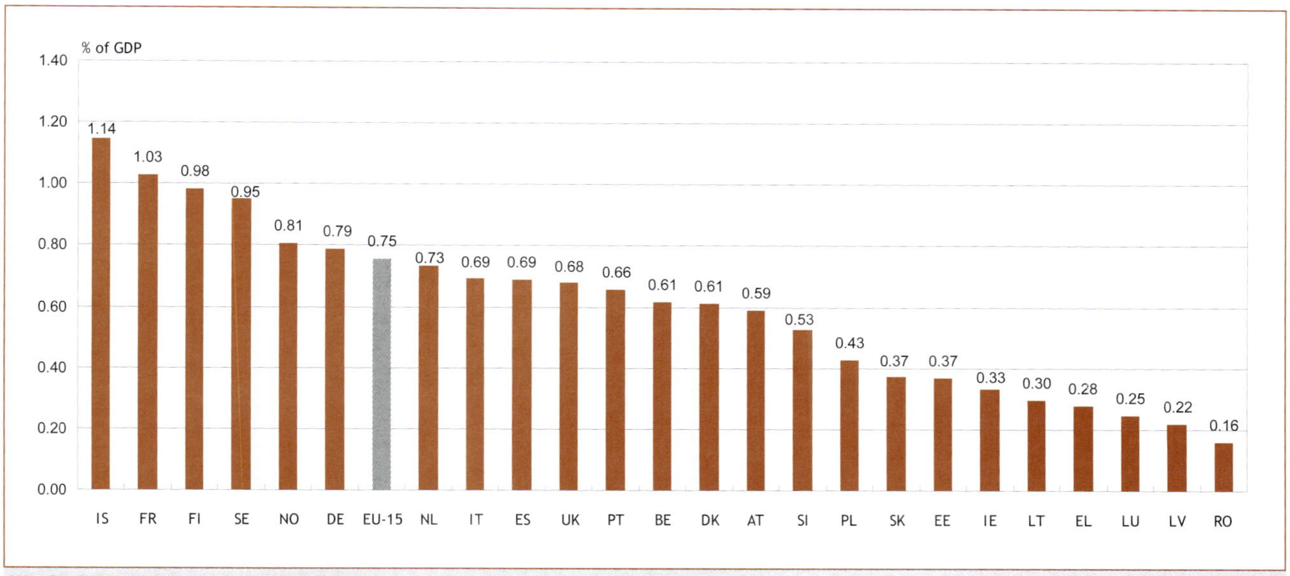

NB: CZ, CY, HU, MT, BG and TR are not included as there are no data available for these countries.
(1) **Exceptions to the reference year 2003**
DK and NL: 2004;
EU-15, EL and FR: 2002;
IE, IT, UK, LV, LT and RO: 2001;
ES, EE, SI and SK: 2000;
PL: 1999.
(2) EU-15, BE, DK, DE, EL, ES, FR, IE, IT, LU, NL, AT, PT, FI, SE, IS and NO: provisional data.
EU-15: Eurostat estimate.

Source: Eurostat.

Figure 1.6.

GBAORD as a % of total general government expenditure
EU-15, Iceland and Norway
2002 (1, 2)

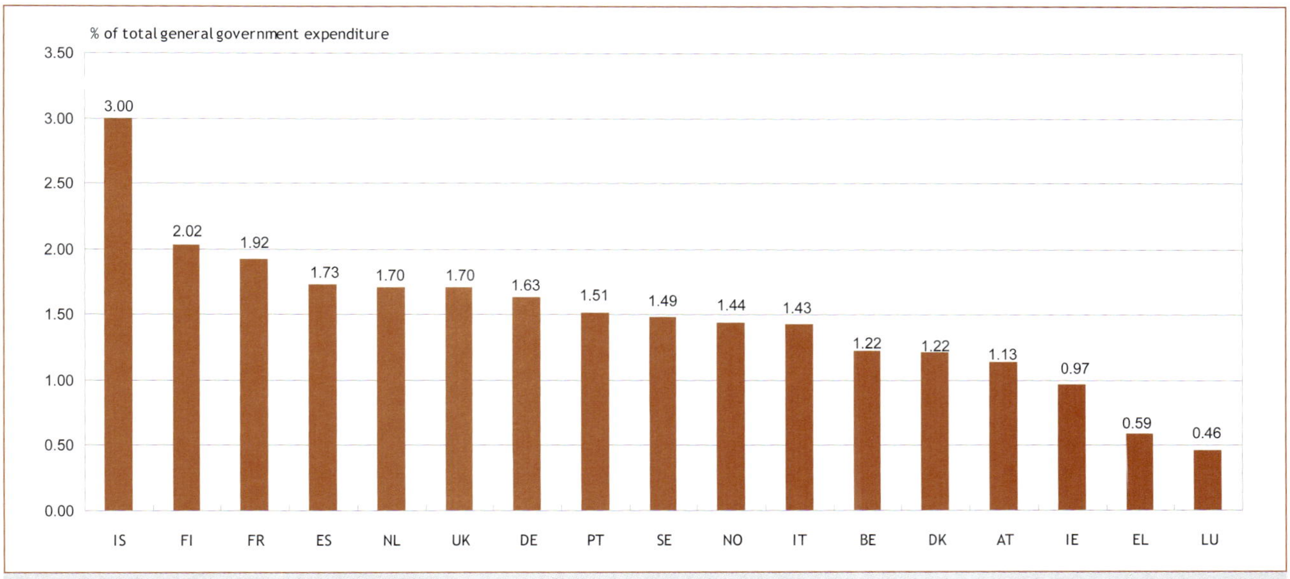

(1) **Exceptions to the reference year 2002**
IT, UK, IS and NO: 2001;
ES: 2000.
(2) EL, FR, IE, IT, AT, FI and SE: provisional data.

Source: Eurostat.

Figure 1.7. shows that in the EU, GBAORD expressed in real terms decreased between 1992 and 1997 at an annual average growth rate of -1.3%. Between 1997 and 2002, GBAORD increased at an annual average growth rate of 2.7%, while the annual average growth rate of GDP was 2.4%.

However, large differences exist across Member States. Between 1992 and 1997, four countries – Germany, France, Italy and Sweden – saw their GBAORD significantly decreased. GBAORD in Norway and in the United Kingdom remained quite stable. All the other countries saw their GBAORD increase. For example, in Greece and Iceland, the annual average growth rate of GBAORD was above 12% between 1992 and 1997.

Therefore, GBAORD decreased at the European level between 1992 and 1997 which was mainly due to the stagnation or decline in GBAORD by the large EU countries – Germany, France, Italy and United Kingdom.

With the exception of Denmark for the 1997-2004 period, between 1997 and 2002, GBAORD of each Member State increased, but large differences still exist between countries.

Ten countries had an annual average growth rate for GBAORD higher than that of their GDP for the same period. These were Belgium, Spain, France, Ireland, Italy, Luxembourg, the Netherlands, Portugal, Iceland and Norway. Among them, GBAORD in Spain, Ireland, Luxembourg and Iceland grew at annual average growth rates of above 12%.

The annual average growth rate of GBAORD in Norway between 1997 and 2002 was approximately equal to the GDP rate. In other countries, the annual average growth rate of GBAORD was lower than GDP rate even if it was positive. In fact, whilst GBAORD increased more than GDP at the European level between 1997 and 2002, it was mainly due to the large rise in GBAORD by Spain, Ireland, Italy, Luxembourg, Portugal and Iceland.

Figure 1.7. **Annual average real growth rates for GBAORD and GDP in %**
EU-15, Iceland and Norway
1992 to 1997 and 1997 to 2002 (1, 2)

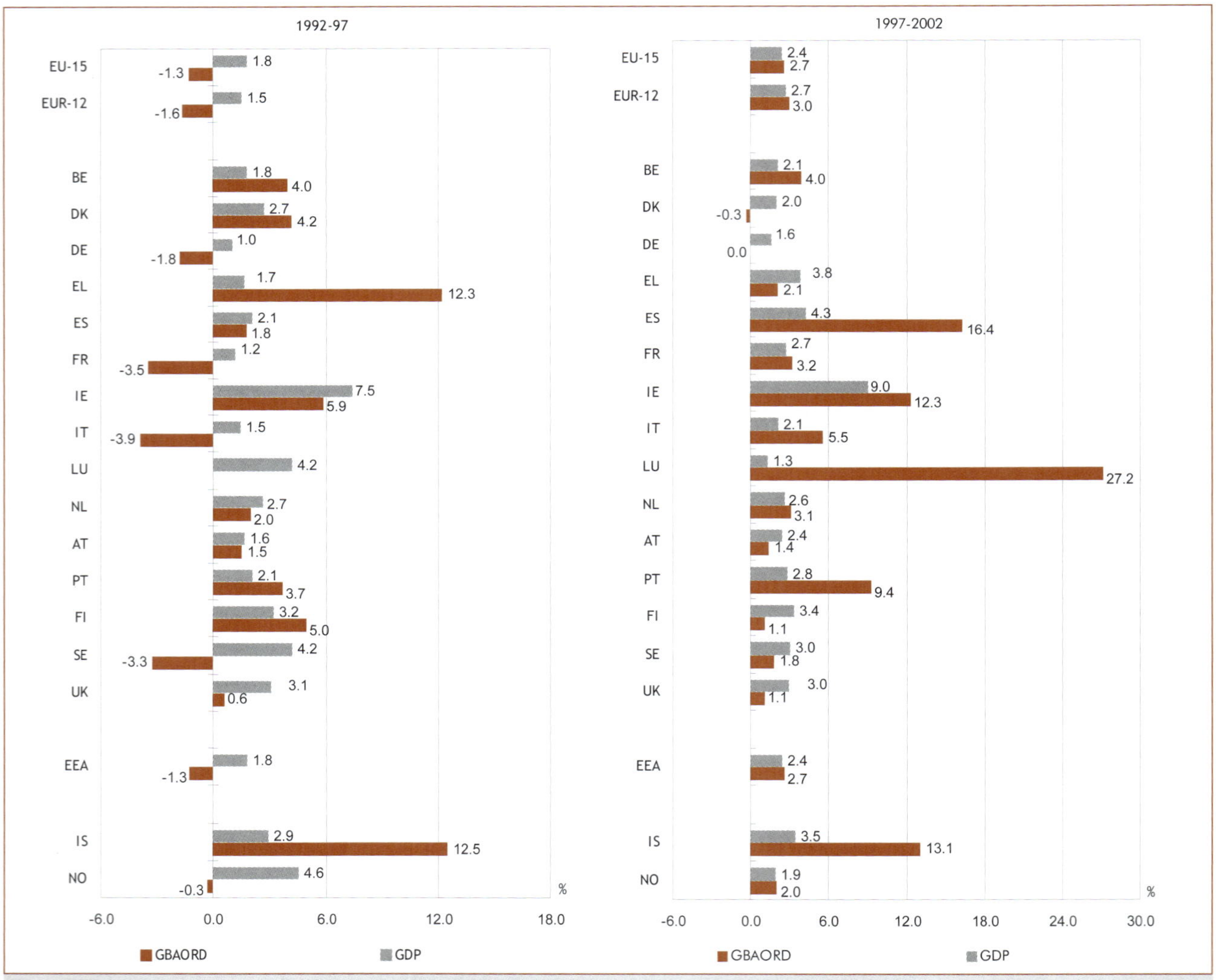

NB: Annual average growth rates — AAGR — are calculated in constant 1995 PPS.
(1) **Exceptions to the reference period 1997-2002** — ES: 1997-2000; IE, IT, UK: 1997-2001; DK: 1997-2004; SE: 1998-2002; LU: 2000-2002.
Exception to the reference period 1992-97 — SE: 1998-2002.
(2) **GBAORD data**
DK: 2004; EL, FR, AT, FI, SE, IS and NO: 2002; IE and IT: 2001; ES: 2000; SE 1998: provisional data.
EU-15 and EUR-12: Eurostat estimates.

Source: Eurostat.

GBAORD by socio-economic objective

The distribution of GBAORD by socio-economic objective varies across Member States

As previously stated, GBAORD is broken down by socio-economic objectives on the basis of the NABS classification.

The main grouped socio-economic objective within the EU in 2002 was 'Research financed from General University Funds (GUF)' as it accounted for 32.4% of total GBAORD — Figures 1.8. and 1.9.

Within the EU, another important objective was 'Technological objectives' — Figure 1.9. These include objectives such as 'Production, distribution and rational utilisation of energy', 'Industrial production and technology' and 'Exploration and exploitation of space'. The other priorities at the European level were mainly 'Non-oriented research' and 'Defence' which claim around 15% of total GBAORD or a little more than 8.4 million of constant 1995 PPS — Figure 1.9.

During the period 1992-2002, the key objective in the EU 'Research financed from General University Funds (GUF)' rose significantly from 15.4 to 19.2 thousand million of 1995 constant PPS. This growth was primarily at the expense of 'Technology' and 'Defence' which fell from 13.2 to 11.7 thousand million and from 10.9 to 8.9 thousand million of 1995 constant PPS respectively.

The 'Human and social' and 'Non-oriented research' objectives also rose over this period. Finally, 'Other civil research' and 'Agricultural production and technology', which amounted to around 0.8 and 1.8 thousand million constant 1995 PPS, decreased during the 1992-2002 period.

Figure 1.8. **Distribution of GBAORD by socio-economic objective in %**
EU-15
1992, 1997 and 2002 (1, 2)

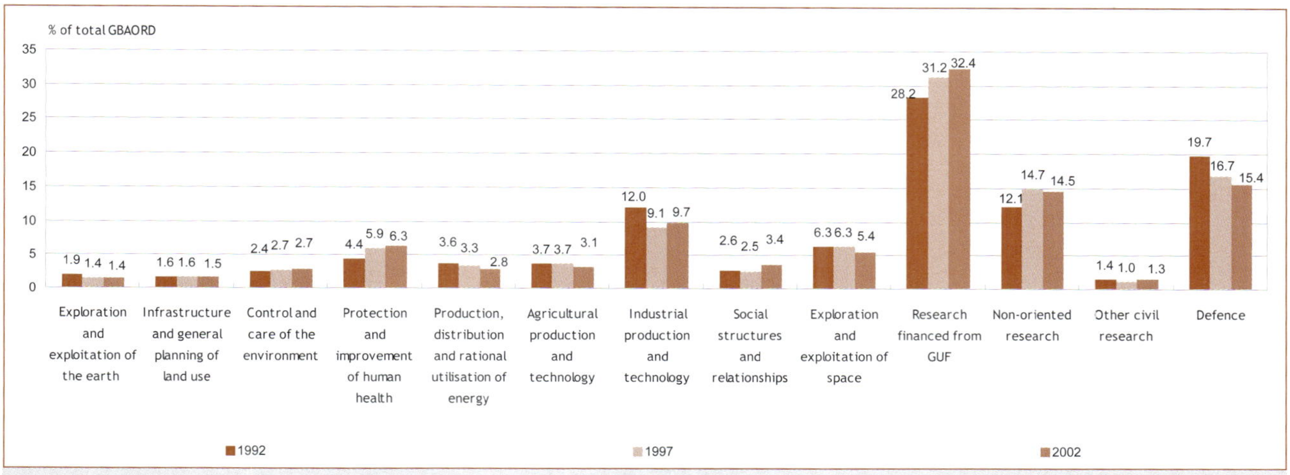

(1) EU-15 excludes LU.
(2) 2002: Eurostat estimates.

Source: Eurostat.

Figure 1.9. **GBAORD by grouped socio-economic objective in millions of constant 1995 PPS**
EU-15
1992 to 2002 (1, 2)

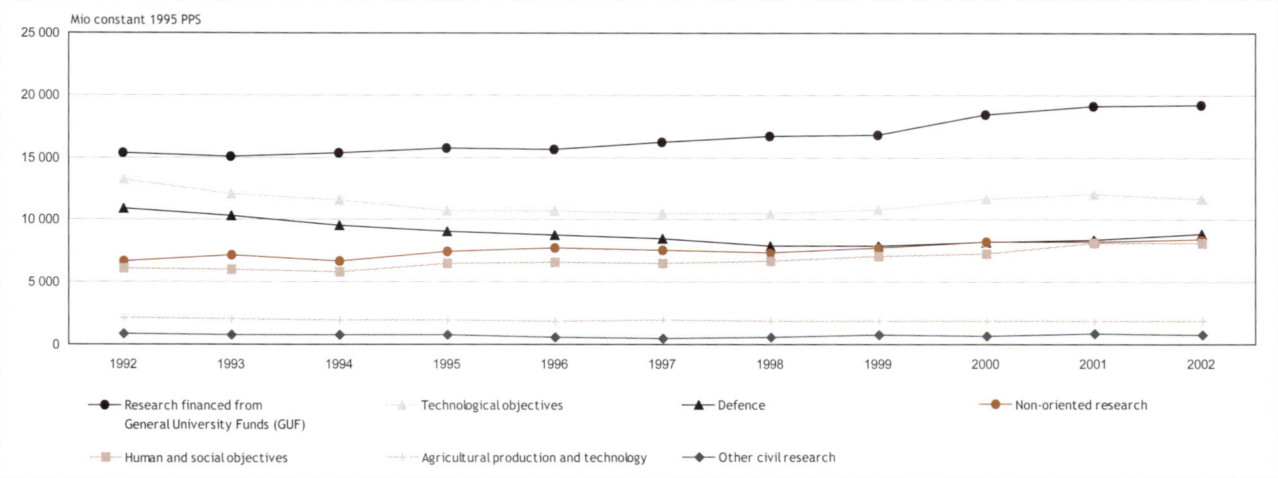

NB: See composition of each group in methodological notes starting on page 150.
(1) EU-15 excludes LU.
(2) 1995-96 and 1999-2002: Eurostat estimates.

Source: Eurostat.

It could be seen that whilst GBAORD — calculated on the basis of constant PPS — in the EU decreased during the 1992-97 period, -1.3% per annum, it grew at an annual average growth rate of 2.7% between 1997 and 2002. However, differences appear not only among countries, but also across socio-economic objectives — Figure 1.10.

In the EU, only four socio-economic objectives increased between 1992 and 1997, namely 'Control and care of the environment', 'Protection and improvement of human health', 'Research financed from General University Funds (GUF)' and 'Non-oriented research'.

All the other socio-economic objectives decreased during the period 1992-97. For example, the objectives 'Exploration and exploitation of the earth', 'Industrial production and technology' and 'Other civil research' decreased at annual average growth rates of -7.2%, -6.5% and -8.8% respectively.

On the contrary, between 1997 and 2002, only three objectives decreased (in real terms). These were 'Production, distribution and rational utilisation of energy', 'Agricultural production and technology' and 'Exploration and exploitation of space', with annual average growth rates of -0.6%, -1.2% and -0.4%, respectively. The four objectives 'Exploration and exploitation of the earth', 'Infrastructure and general planning of land use', 'Non-oriented research' and 'Defence' increased between 1997 and 2002 but at annual average growth rates below or equal to rates registered by total GBAORD (2.7%).

The socio-economic objectives that increased the most during the 1997-2002 period were 'Social structure and relationships' and 'Other civil research' with annual average growth rates above 8%. The rest of the objectives increased at annual average growth rates comprised between 3.3% and 4.5%.

Figure 1.10. **Annual average real growth rates of GBAORD by socio-economic objective in %**
EU-15
1992 to 1997 and 1997 to 2002 (1)

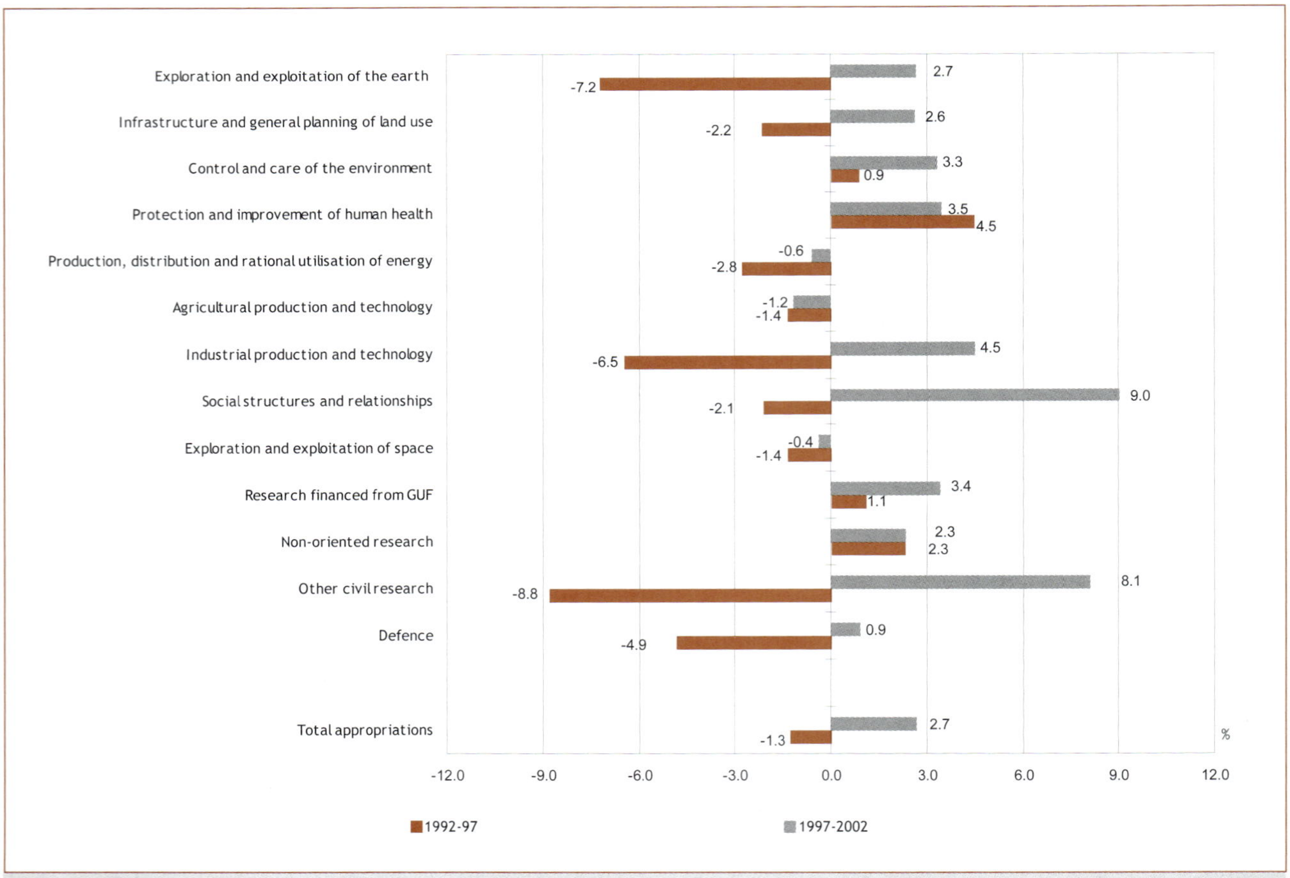

NB: Annual average growth rates — AAGR — are calculated in constant 1995 PPS.
(1) 1997 and 2002: Eurostat estimates.

Source: Eurostat.

However, the distribution across socio-economic objectives and their evolution show large differences among Member States. For example, the objective 'Defence' at the European level was the second priority in 2002 with 15.4% of total GBAORD. However, if 'Defence' represents a substantial part of the EU's total GBAORD, this is mainly due to the contribution of certain countries. In fact, 'Defence' is the main or one of the main objectives for Spain (30.2%), France (24.2%), Sweden (22.2%) and the United Kingdom (34.9%), but it represents less than 7% of national total GBAORD for the rest of the countries — Table 1.2.

For certain countries, such as Ireland, Iceland and Portugal, the objective 'Agricultural production and technology' is quite significant due to the importance of fishing activities in Iceland and agricultural activities in Ireland and Portugal. The 'Industrial production and technology' objective represented a noteworthy part of GBAORD in Belgium, Germany, Spain, Ireland, Italy, Portugal and Finland. In Iceland, the main socio-economic objective is 'Social structure and relationships'.

For the research funded by the Commission of the European Communities — CEC, 'Industrial production and technology' is the main objective — 1999 data, as it accounts for one third of the total budget. This is followed by 'Production, distribution and rational utilisation of energy' (15.2%).

Table 1.2.

Distribution of GBAORD by socio-economic objective in %
EU-15, Iceland and Norway
2003 (1, 2)

	Socio-economic objectives	EU-15	CEC	BE	DK	DE	EL	ES	FR	IE	IT	LU	NL	AT	PT	FI	SE	UK	IS	NO
1.	Exploration and exploitation of the earth	1.4 s	1.9	0.7	0.6	1.7	4.1	2.0	0.7	3.0	1.9	:	0.4	2.3	1.9	1.0	0.3	1.6	-	2.0
2.	Infrastructure and general planning of land use	1.5 s	6.9	1.4	1.4	1.8	2.6	0.6	0.6	2.7	0.4	:	4.7	2.2	4.9	0.3	2.6	1.5	7.7	2.2
3.	Control and care of the environment	2.7 s	7.6	2.3	1.9	3.1	3.7	2.7	2.9	1.7	2.3	:	2.6	1.7	3.4	2.0	1.5	1.9	0.3	2.5
4.	Protection and improvement of human health	6.3 s	8.0	1.8	1.6	4.1	6.4	4.8	5.8	4.0	7.0	:	3.2	2.8	7.6	6.5	1.0	14.9	9.1	7.6
5.	Production, distribution and rational utilisation of energy	2.8 s	15.2	2.1	1.3	2.9	1.9	3.6	3.7	-	3.6	:	3.6	0.6	1.3	4.4	2.9	0.5	2.3	1.8
6.	Agricultural production and technology	3.1 s	5.3	2.1	7.8	2.0	6.7	4.2	2.1	23.3	1.8	:	4.1	2.6	12.3	6.1	2.7	3.9	20.9	9.1
7.	Industrial production and technology	9.7 s	33.8	31.3	6.8	12.4	6.8	15.8	5.9	15.3	14.8	:	12.0	7.3	17.0	26.6	5.4	3.9	2.4	8.6
8.	Social structures and relationships	3.4 s	3.8	4.9	5.9	4.7	4.8	0.6	0.9	4.0	4.4	:	2.8	1.7	3.8	7.4	6.2	4.0	41.3	7.2
9.	Exploration and exploitation of space	5.4 s	0.7	8.9	2.3	4.9	0.1	5.5	8.9	-	7.3	:	2.5	0.1	0.5	1.9	0.6	2.1	-	2.0
10.	Research financed from General University Funds (GUF)	32.4 s	-	18.2	47.1	39.3	50.6	21.4	23.0	18.3	43.7	:	47.6	65.5	33.5	27.3	38.0	21.7	-	40.2
11	Non-oriented research	14.5 s	6.5	22.9	22.0	16.6	11.0	7.3	19.7	27.6	8.8	:	10.7	13.1	9.9	13.5	16.7	13.5	16.0	12.6
12.	Other civil research	1.3 s	10.2	3.1	-	-0.2	0.3	1.2	1.5	-		:	4.0	0.0	2.0	-	-	0.3	-	-
13.	Defence	15.4 s	-	0.4	1.3	6.7	0.9	30.2	24.2	-	4.0	:	1.7	0.0	2.0	2.9	22.2	30.3	-	4.2
	Total civil appropriations	84.6 s	100.0	99.6	98.7	93.3	99.1	69.8	75.8	100	96.0	100	98.3	100	98.0	97.1	77.8	69.7	100	95.8
	Total appropriations in Mio EUR	69 163 s	3 148	1 650	1 202	16 930	392	4 187	15 609	378	8 441	57	3 476	1 303	880	1 417	2 522	100	110	1 652

(1) **Exceptions to reference year 2003**
DK and NL: 2004; EU-15, EL and FR: 2002; IE, IT and UK: 2001; ES: 2000; CEC: 1999.
(2) Provisional data, except UK 2001 final.

Source: Eurostat.

As shown in Table 1.3., the priorities for most Candidate Countries were, in 2000, the objectives 'Other civil research', 'Non-oriented research' and 'Industrial production and technology'. In these countries, the objective 'Defence' is insignificant compared to the European average (15.4%).

Table 1.4. shows annual average real growth rates recorded by socio-economic objective between 1997 and 2002 for European Union Member States, Iceland and Norway. Once again, the evolution by socio-economic objective at the European level shows large differences among countries.

At the European level, total appropriations increased at an annual average growth rate of 2.7%, but total civil appropriations grew faster at 3.0% per annum. The two objectives that recorded the largest annual average growth rates were 'Social structures and relationships' and 'Other civil research'.

On the contrary, the objectives that decreased the most were 'Agricultural production and technology' (-1.2%), 'Production, distribution and rational utilisation of energy' (-0.6%) and 'Exploration and exploitation of space' (-0.4%). However, at the national level, the situation varies considerably.

Table 1.3.

Distribution of GBAORD by socio-economic objective in %
Candidates Countries
2000 (1, 2)

Socio-economic objectives	EE	LV	LT	PL	SI	SK	RO
1. Exploration and exploitation of the earth	:	0.9	1.6	:	0.7	:	3.0
2. Infrastructure and general planning of land use	:	0.3	5.2	:	1.5	1.3	10.4
3. Control and care of the environment	:	2.4	5.2	:	1.5	1.3	3.7
4. Protection and improvement of human health	:	11.2	10.3	:	1.5	4.9	4.0
5. Production, distribution and rational utilisation of energy	:	2.0	0.9	:	1.0	1.5	3.3
6. Agricultural production and technology	:	13.4	5.4	:	4.8	13.2	12.7
7. Industrial production and technology	:	16.5	15.6	:	17.8	11.2	31.5
8. Social structures and relationships	:	5.9	8.7	:	2.1	10.9	1.7
9. Exploration and exploitation of space	:	1.3	:	:	0.0	:	2.1
10. Research financed from General University Funds (GUF)	:	:	:	:	4.2	18.2	:
11. Non-oriented research	:	20.7	:	:	64.9	29.6	23.5
12. Other civil research	:	24.6	47.0	:	:	7.7	2.8
13. Defence	:	0.7	0.1	:	0.1	:	1.4
Total civil appropriations	:	99.3	99.9	:	99.9	:	98.6
Total appropriations in Mio EUR	**20**	**16**	**39**	**623**	**107**	**79**	**72**

NB: CZ, CY, HU, MT, BG and TR are not included as there are not data available for these countries.

(1) **Exceptions to the reference year 2000**
PL: 1999;
LT and RO: 2001.

(2) EE: estimated value.

Source: Eurostat.

For example, in Spain, Sweden, Portugal, Denmark, and Italy the objective 'Defence' increased, between 1997 and 2002 at a faster rate than total GBAORD. But it decreased between -2.5% and -10.6% per annum in Norway, Greece, the United Kingdom, Belgium, the Netherlands and Germany.

Whilst 'Social structures and relationships' was the most dynamic socio-economic objective in the EU, as it increased at 9.0% per annum, rates by country in the EEA ranged from 20.7% in the United Kingdom to -4.0% in Denmark. Government appropriations on this objective also grew above the average in Germany, Belgium, the Netherlands, Iceland and Spain.

The second fastest growing objective in the EU, 'Other civil research' (8.1%), was most dynamic in Spain (13.3% per annum) and decreased the most in Austria (-11.4%). Rates for 'Industrial production and technology' varied from 23.6% in the United Kingdom to -14.1% in Sweden.

Although the objective 'Agricultural production and technology' decreased overall in the EU, all Spain, Iceland and Sweden recorded growth rates above 10%; Portugal, Ireland, Norway, the Netherlands and Italy also recorded positive annual average growth rates.

Table 1.4.

Annual average real growth rates of GBAORD by socio-economic objective in %
EU-15, Iceland and Norway
1997 to 2004 provisional ([1, 2])

	Socio-economic objectives	EU-15	BE	DK	DE	EL	ES	FR	IE	IT	LU	NL	AT	PT	FI	SE	UK	IS	NO
1.	Exploration and exploitation of the earth	2.7	2.7	-10.3	-3.2	-0.9	12.9	0.5	87.0	14.4	:	0.9	4.9	2.4	-1.0	-21.6	4.3	-	-3.6
2.	Infrastructure and general planning of land use	2.6	13.1	-1.0	1.4	3.2	36.3	-0.2	13.4	3.0	:	7.6	-0.2	23.1	-26.9	-8.5	-2.1	19.3	2.8
3.	Control and care of the environment	3.3	5.8	-6.2	-2.6	4.1	24.9	10.2	7.5	3.8	:	2.9	-4.5	5.1	0.1	7.6	-3.7	21.7	-1.6
4.	Protection and improvement of human health	3.5	5.4	-0.1	4.5	11.1	15.4	3.8	24.7	5.4	:	12.8	-4.0	18.3	-0.2	-1.8	2.1	27.4	3.6
5.	Production, distribution and rational utilisation of energy	-0.6	0.0	-10.0	-3.1	1.4	9.6	-2.1	-	2.7	:	2.9	-10.9	8.1	2.5	-6.7	-5.0	9.6	-0.1
6.	Agricultural production and technology	-1.2	-7.3	-1.9	-5.9	-3.1	18.3	-8.2	6.1	0.2	:	3.2	-3.4	8.9	-1.4	10.7	-2.8	11.2	5.7
7.	Industrial production and technology	4.5	11.3	-6.3	-0.3	-3.9	12.9	8.0	-4.9	19.6	:	2.4	4.7	18.6	1.4	-14.1	23.6	22.1	-7.0
8.	Social structures and relationships	9.0	13.0	-4.0	13.9	8.1	9.1	2.0	4.9	7.4	:	10.3	-0.6	7.6	6.0	4.5	20.7	9.7	0.7
9.	Exploration and exploitation of space	-0.4	-0.9	-3.5	1.4	-23.1	10.2	-2.1	-	0.8	:	1.7	55.2	19.3	-3.6	5.6	-6.3	-	-1.6
10.	Research financed from General University Funds (GUF)	3.4	0.6	3.5	0.7	1.9	5.4	9.2	6.5	4.6	:	3.5	1.7	3.7	1.6	3.4	6.6	-	3.2
11.	Non-oriented research	2.3	3.5	-1.2	1.9	6.7	14.0	2.9	72.2	0.0	:	1.3	1.9	16.3	3.4	-	5.5	16.3	11.7
12.	Other civil research	8.1	-5.0	-	-1.2	11.0	13.3	-6.3	-	-	:	1.0	-11.4	-5.8	-	-	-4.6	-	-
13.	Defence	0.9	-6.9	15.8	-10.6	-4.0	34.4	2.4	-	3.2	:	-9.0	-	16.4	1.4	33.0	-5.2	-	-2.5
	Total civil appropriations	3.0	4.0	-0.4	0.9	2.1	11.0	3.5	12.3	5.6	27.2	3.4	1.4	9.2	1.1	2.6	4.6	13.1	2.2
	Total appropriations	**2.7**	**4.0**	**-0.3**	**0.0**	**2.1**	**16.4**	**3.2**	**12.3**	**5.5**	**27.2**	**3.1**	**1.4**	**9.4**	**1.1**	**5.9**	**1.1**	**13.1**	**2.0**

NB: Annual average growth rates — AAGR — are calculated in constant 1995 PPS.

(1) **Exceptions to the reference period 1997-2002**
ES: 1997-2000; IE, IT and UK: 1997-2001; SE: 1998-2002; LU: 2000-2002.

(2) **Provisional data**
EL, FR, AT, FI, SE, IS and NO: 2002; IE and IT: 2001; ES: 2000; SE: 1998.
EU-15: Eurostat estimates.

Source: Eurostat.

2.1. Introduction

R&D activities are often considered as a main drive for economic development, innovation and growth. They comprise creative work undertaken on a systematic basis in order to increase the stock of knowledge, including knowledge of man, culture and society, and the use of this stock of knowledge to devise new applications. The basic statistical variables are R&D expenditure and number of R&D personnel — see Chapter 3, which are measured both at national and regional levels.

R&D expenditure corresponds to the measurement of 'intramural' expenditure, i.e. all expenditure for R&D performed within a statistical unit or sector of the economy during a specific period, whatever the source of funds ([1]). Intramural R&D expenditure is broken down by institutional sector, i.e. by sector engaged in R&D.

In this publication, five sectors are used to calculate indicators of R&D activity:

- the Business enterprise sector — BES,
- the Government sector — GOV,
- the Higher education sector — HES,
- the Private non-profit sector — PNP — and
- all sectors which corresponds to the sum of the four previous sectors.

However, given the minor role played by the PNP sector in all countries except in Portugal, this sector has not been systematically included in all the analyses in this chapter.

Two manuals are used as methodological references for R&D surveys:

- the *Frascati Manual* and
- the *Regional Manual* ([2]).

They provide a model for obtaining comparable statistics between countries.

This chapter presents the key indicators for R&D expenditure as well as the main trends for the last decade. In an effort to respond the increasing demand of R&D expenditure statistics by policy makers, new indicators prepared for this Panorama edition, such as the source of funds or data by fields of science, are also included. The complete time series are available at Eurostat's reference database *NewCronos*. Certain data series originating from the OECD have also been utilised in this chapter ([3]).

In addition to data for the EU countries, Iceland, Norway, Japan and the United States, for the first time, this chapter also gathers R&D expenditure data for Candidate Countries, including an Acceding Countries aggregate — ACC. In order to facilitate the reading, in the text Candidate Countries will refer to both the ten Acceding Countries and the three Candidate Countries.

This chapter is divided into three sections:

- Firstly, it focuses on R&D expenditure in the EU, the Acceding Countries — ACC, Japan and the United States.
- Secondly, the main trends at the national level are highlighted, by looking at the performance of the EU Member States, Candidate Countries, Iceland and Norway.
- Finally, R&D expenditure at the regional level is analysed, focusing on the regions of the EU countries, Iceland and Norway.

Although the regional analysis is carried out at the NUTS 2 level, other levels of NUTS are sometimes used for particular countries, this being specified in each case by means of a footnote. Readers should also notice that according to the NUTS classification, for Denmark and Luxembourg the entire national territory is considered as a NUTS 0, 1 or 2 region and therefore Denmark and Luxembourg may appear in rankings at the NUTS 2 level.

The analysis refers to the period 1993-2002. However, the time series do not cover the same period for all countries. In general, when data for the year 2002 are not available for a particular country, the latest year available is presented.

(1) ***Standard method for surveys on R&D and experimental development — Frascati Manual,*** OECD 2002, paragraph 358.

(2) ***The regional dimension of R&D statistics and of innovation — Regional Manual,*** Eurostat, 1996.

(3) ***Main Science and Technology Indicators*** — MSTI, OECD.
Data for Japan and the United States uses MSTI 2002/2 data.
Data for R&D expenditure by source of funds uses MSTI 2003/1 data.

2.2. R&D expenditure in the EU, ACC, Japan and the United States

EU R&D expenditure increased in 2002, but the gap with the United States and Japan remains

In 2002 R&D expenditure as a share of GDP in the EU increased to 1.99%. However, the gap with regard to R&D expenditure in Japan (2.98% in 2000) and the United States (2.80%) remained unchanged. The level reached by the Acceding Countries – ACC – was stable during the 1998-2000 period, reaching a share of 0.83% in 2000, being still far below the levels of the previously mentioned triumvirate. Since 1997, the trend for the R&D intensity for the EU, Japan and the United States was positive even if the United States showed a small decrease of 0.02 percentage points in 2002. Nevertheless, the evolution of the trend for these three blocks was made at different rhythms. Japan gained 0.15 percentage points between 1997 and 2000, ahead of the United States – 0.14 percentage points, and the EU – 0.08 percentage points, over the same period. If the gap between the EU and the United States was enlarged from 0.72 percentage points in 1998 to 0.81 in 2002, it decreased slightly from 1.06 percentage points in 1998 to 1.03 in 2000 when compared with Japan – Figure 2.1.

The EU devoted EUR 182 thousand million in 2002 to R&D expenditure as compared to EUR 309 thousand million for the United States and EUR 154 thousand million for Japan in 2000. The growth displayed in nominal terms between 1996 and 2002 was 40% for the EU, 50% for Japan (1996-2000) whereas the United States doubled its amount of R&D expenditure over the last seven years – Figures 2.2. and 2.3. Measured in PPS at 1995 constant prices, the growth and the differences between the United States and the two other blocks are more moderate. For the same periods, the growth for the United States was 33% as against 24% for the EU and 12% for Japan. As shown in Figure 2.3., the gaps for R&D expenditure between the United States on the one hand and the EU and Japan on the other hand, have expanded in real terms since 1994. In 1994, there was a difference of EUR 43 thousand million of constant 1995 PPS between the EU and the United States, which by 2002 doubled and reached to EUR 89 thousand million.

Most of R&D expenditure is carried out in the Business enterprise sector – BES. The BES accounts for 65% of R&D expenditure in the EU, which is below the percentages seen in the United States (73%), and Japan (71%). Between 1996 and 2002, this ratio remained stable for both Japan and the United States but increased by 2 percentage points for the EU – Figure 2.2. This common stability of the share of R&D expenditure is also a general rule for the Higher education sector – HES. The main changes occurred in the Government sector – GOV – where the proportion of R&D expenditure decreased in the EU (3 percentage points) and the United States (1 percentage points) but increased in Japan (1 percentage point).

Figure 2.1.

R&D expenditure as a % of GDP, all sectors
EU-15, ACC, Japan and the United States
1993 to 2002

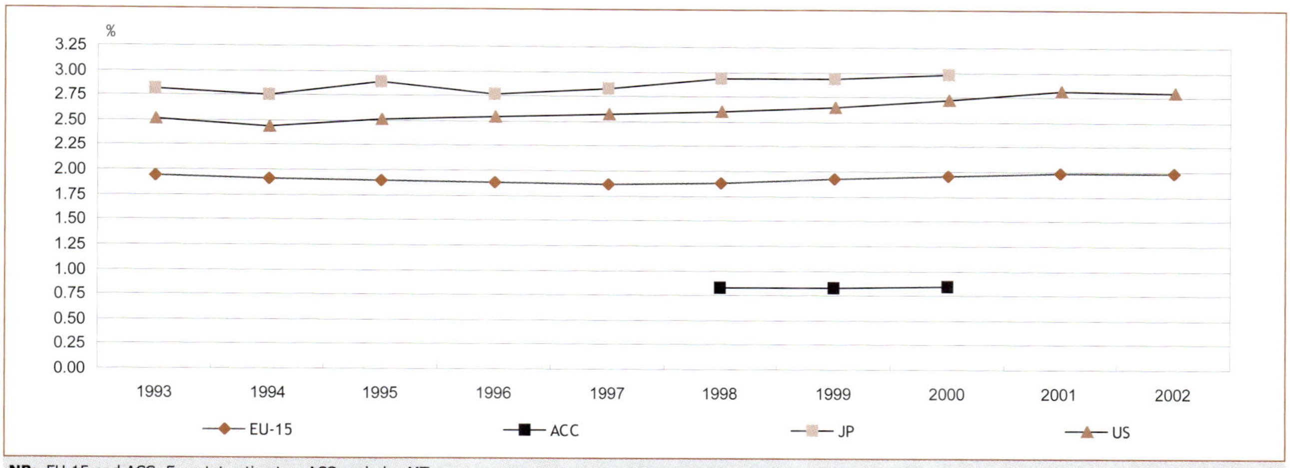

NB: EU-15 and ACC: Eurostat estimates. ACC excludes MT.
JP 1996: break in series; US 2002: estimated value.

Sources: Eurostat; OECD.

Figure 2.2.

R&D expenditure in EUR thousand million, by institutional sector
EU-15, Japan and the United States
1996 and 2002

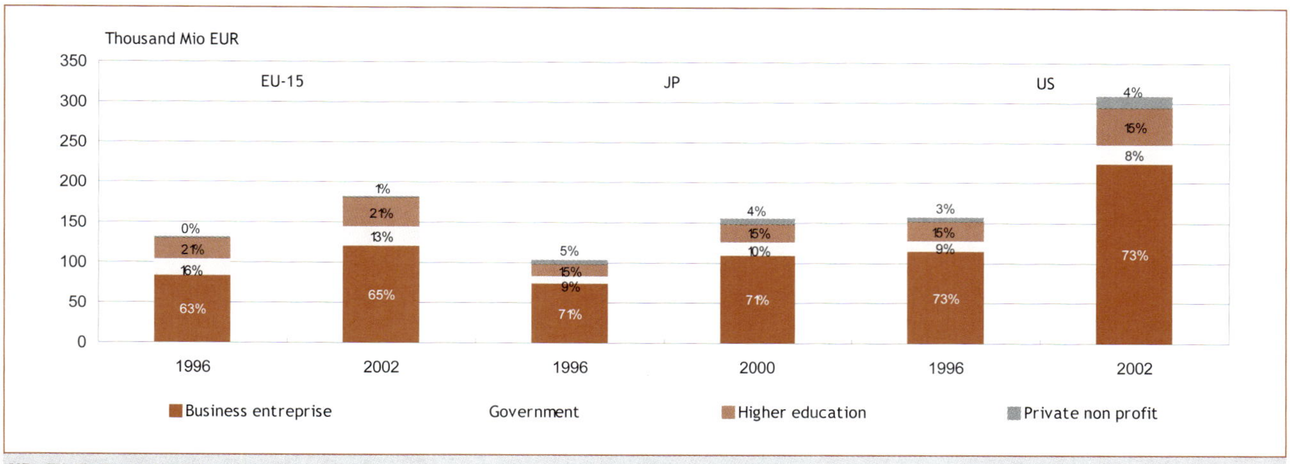

NB: EU-15: Eurostat estimates.
US 2002: estimated values.

Sources: Eurostat; OECD.

Figure 2.3.

R&D expenditure in EUR thousand million and thousand million constant 1995 PPS, all sectors
EU-15, Japan and the United States
1993 to 2002

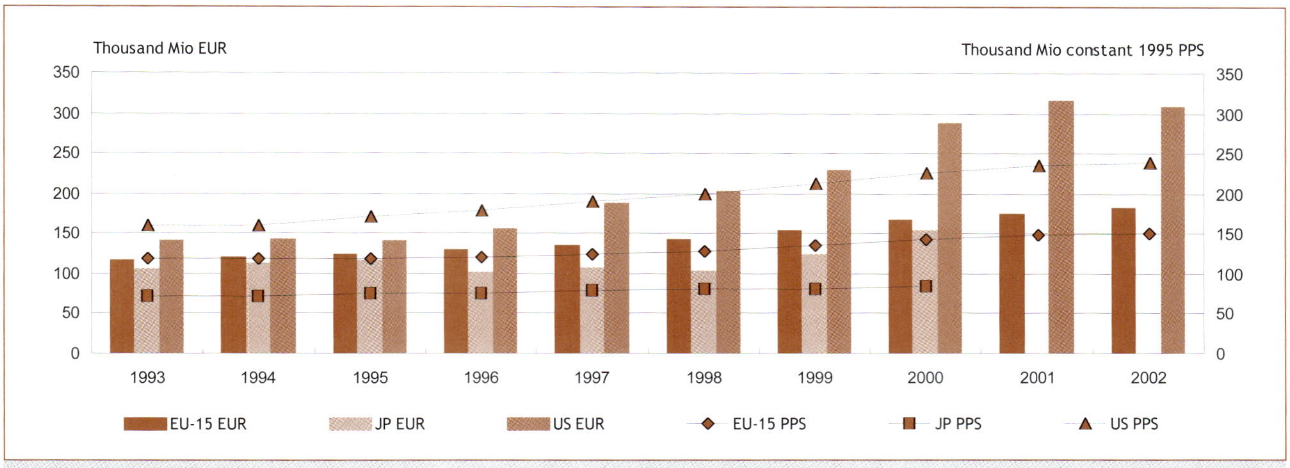

NB: EU-15: Eurostat estimates.
JP 1996: break in series; US 2002: estimated values.

Sources: Eurostat; OECD.

2.3. R&D expenditure at the national level

R&D intensity

Sweden, Finland and Iceland are the top European countries in terms of R&D expenditure as a percentage of GDP. In absolute terms, Germany is leading

According to the latest data available, the leading EEA countries in terms of R&D intensity are Sweden, Finland and Iceland with 4.27% (2001 data), 3.49% and 3.11% – both 2002 estimated data – of the GDP devoted to R&D expenditure respectively. With ratios above 3%, they come ahead of Japan and the United States. In particular, the upward trend over the last years for Sweden and Iceland was noticeable as R&D intensity increased more than 0.5 percentage points since 1999 – data not available in Table 2.1., see ***NewCronos***. Other EU countries with the R&D intensity rates above the EU average are Germany (2.49%), Denmark (2.40%), France (2.20%) and Belgium (2.17%) – all 2001 data, except France 2002. R&D intensity in France and the United Kingdom in 2002 decreased compared to 2001.

In the Business enterprise sector – BES, Sweden (3.31%, 2001 data) and Finland (2.47%, 2002 estimated data) also had the highest R&D intensity, whereas the highest increase compared to the previous year was retained by Denmark (0.14% percentage points between 2000 and 2001). In the Government sector – GOV, the highest ratios were registered by Iceland (0.76%), France and Finland (0.37% each) – all 2002 estimated data, whereas in the Higher education sector – HES – Sweden (0.83%, 2001 data) and Finland (0.65%, 2002 estimated data) were leading.

The R&D intensity in the Candidate Countries is on average below the levels observed for the EU. Although R&D intensity is above 1.30% for the Czech Republic and Slovenia, the rest of the Candidate Countries retained figures below 1% in 2001. Slovenia (0.86%) and Czech Republic (0.80%) showed particularly high figures in the BES. By contrast to the EU countries, where the BES comes as first sector in terms of R&D intensity, in Cyprus, Bulgaria, and, to a lesser extent, Lithuania, the GOV sector comes first, whereas in Estonia, Latvia and Turkey, HES is the most important sector – Table 2.1.

Table 2.1.

R&D expenditure as a % of GDP, by institutional sector
EU-15, Candidate Countries, Iceland, Norway, Japan and the United States
2000 to 2002 (1)

	All sectors			Business enterprise			Government			Higher education		
	2000	2001	2002	2000	2001	2002	2000	2001	2002	2000	2001	2002
EU-15	1.95 s	1.98 s	1.99 s	1.27 s	1.30 s	1.30 s	0.26 s	0.25 s	0.26 s	0.40 s	0.41 s	0.42 s
BE	2.04 ep	2.17 ep	:	1.48 ep	1.60 ep	1.63 ep	0.13 ep	0.13 ep	:	0.40 ep	0.41 ep	:
DK	2.26 er	2.40	:	1.51 er	1.65	:	0.28 r	0.28	:	0.45 r	0.45	:
DE	2.49 e	2.49 e	:	1.75 e	1.76 e	:	0.34	0.33 e	:	0.40	0.40 e	:
EL	0.67 e	:	:	0.19 r	:	:	0.15	:	:	0.33	:	:
ES	0.94 er	0.96 r	:	0.50 er	0.50 r	:	0.15 r	0.15 r	:	0.28 r	0.30 r	:
FR	2.18	2.23 p	2.20 e	1.36	1.41 p	1.37 e	0.38	0.37 p	0.37 e	0.41	0.42 p	0.43 e
IE	1.15	1.17	:	0.83	0.80	:	0.09	0.11 p	:	0.23	0.26	:
IT	1.07	:	:	0.53	0.56	:	0.20	0.22	:	0.33	:	:
LU	1.71 r	:	:	1.58 r	:	:	0.12 r	0.15 r	:	:	0.01 r	:
NL	1.94	:	:	1.11	1.08 p	:	0.27	0.26 p	:	0.57	:	:
AT	1.84 e	1.90	1.94	1.13 r	:	:	0.11 r	:	:	0.53	:	:
PT	:	0.84 e	:	:	0.27 e	:	:	0.18	:	:	0.31	:
FI	3.40	3.40	3.49 f	2.41	2.42	2.47 f	0.38	0.37	0.37 f	0.61	0.61	0.65 f
SE	:	4.27	:	:	3.31	:	:	0.12	:	:	0.83	:
UK	1.85 r	1.89 r	1.84 f	1.21 r	1.28 r	1.19 f	0.22 r	0.18 r	0.22 f	0.38 r	0.41 r	0.41 f
ACC	0.83 s	0.84 s	:	0.40 s	0.39 ps	:	0.25 s	0.24 s	:	0.18 s	0.20 s	:
CZ	1.24	1.33	:	0.78	0.80	:	0.30	0.34	:	0.15	0.19	:
EE	0.75	0.66	:	0.18	0.15	:	0.18	0.15	:	0.38	0.35	:
CY	0.25	0.26	:	0.05	0.06	:	0.12	0.12	:	0.06	0.06	:
LV	0.40	0.48	0.44	0.07	0.19	0.16	0.13	0.11	0.10	0.20	0.18	0.19
LT	0.52	0.60	0.68	0.02	0.13	0.20	0.30	0.25	0.27	0.19	0.22	0.21
HU	0.69	0.80	:	0.28	0.36	:	0.22	0.21	:	0.15	0.19	:
MT	:	:	:	:	:	:	:	:	:	:	:	:
PL	0.75	0.70	:	0.31	0.25	:	0.23	0.23	:	0.21	0.22	:
SI	1.51	1.52	:	0.83	0.86	:	0.43	0.39	:	0.24	0.25	:
SK	0.66	0.67	:	0.42	0.44	:	0.18	0.16	:	0.07	0.06	:
BG	0.56 b	0.52	:	0.12 b	0.11	:	0.41 b	0.36	:	0.03 b	0.05	:
RO	0.40	0.37	:	0.30	0.26	:	0.08	0.07	:	0.03	0.04	:
TR	0.60	:	:	0.20	:	:	0.00	:	:	0.40	:	:
EEA	1.95 s	1.97 s	1.98 s	1.27 s	1.29 s	1.29 s	0.26 s	0.25 s	0.26 s	0.40 s	0.41 s	0.42 s
IS	2.77 e	3.08	3.11 f	1.56 e	1.81	1.78 f	0.76 e	0.62	0.76 f	0.45 e	0.58	0.50 f
NO	:	1.62	:	:	0.97	:	:	0.24	:	:	0.42	:
JP	2.98	:	:	2.11	:	:	0.29	:	:	0.43	:	:
US	2.72	2.82	2.80 e	2.04	2.10	2.04 e	0.18	0.20	0.21 e	0.38	0.40	0.42 e

NB: ACC excludes MT.

(1) **Exceptions to the reference year 2000**
AT – BES, GOV and HES: 1998;
EL: 1999.

Sources: Eurostat; OECD.

R&D expenditure in volume

Whilst larger economies dominate in absolute terms, R&D expenditure grew fastest in Portugal and Ireland

Most R&D in the EU is carried out in Germany (EUR 51.5 thousand million), France (EUR 33.4 thousand million) and the United Kingdom (EUR 30.5 thousand million). These three countries account for almost 2/3 of total R&D expenditure in EU-15. In parallel, the top three countries in terms of R&D intensity, Sweden, Finland and Germany, represent 37% of the EU total. Total R&D expenditure has shown a positive trend during the last years in all EU-15 countries when assessed in current EUR. The annual nominal growth rates ranked from 0.8% in the United Kingdom to 13.1% in Ireland — Table 2.2.

Table 2.2. **R&D expenditure in million EUR, all sectors EU-15, Candidate Countries, Iceland, Norway, Japan and the United States 1999 to 2002**

	1999	2000	2001	2002	AGR (1)	AAGR (2)
EU-15	154 306 s	167 297 s	175 507 s	182 387 s	3.9	5.7
BE	4 618 er	5 040 ep	5 507 ep	:	9.3	9.2
DK	3 406 r	3 892 er	4 265	:	9.6	11.9
DE	48 191 r	50 619 e	51 539 e	:	1.8	3.4
EL	795 e	:	:	:	:	21.1
ES	4 995	5 719 er	6 227 r	:	8.9	11.6
FR	29 529	30 954	32 919 p	33 414 e	1.5	4.2
IE	1 076 e	1 184	1 339	:	13.1	11.6
IT	11 524 r	12 460	:	:	8.1	:
LU	:	364 r	:	:	:	:
NL	7 563 r	7 813	:	:	3.3	3.3
AT	3 656 e	3 806 e	4 031	4 217	4.6	4.9
PT	815 r	:	1 038 e	:	:	12.9
FI	3 879	4 423	4 619	4 873 f	5.5	7.9
SE	8 608	:	10 459	:	:	10.2
UK	25 300	28 788 r	30 255 r	30 501 f	0.8	6.4
ACC	2 580	2 958	3 399	:	14.9	14.8
CZ	641	744	832	:	11.8	13.9
EE	37	37	49	:	31.8	15.5
CY	21	25	27	:	12.2	13.2
LV	25	38	38	:	0.5	23.4
LT	52	73	91	:	24.9	32.9
HU	309	405	548	:	35.2	33.1
MT	:	:	:	:	:	:
PL	1 086	1 197	1 323	:	10.6	10.4
SI	284	297	341	:	14.8	9.7
SK	126	143	149	:	4.5	8.9
BG	69 b	71	71	:	-0.5	1.7
RO	134	149	177	:	18.8	14.7
TR	:	:	:	:	:	:
EEA	156 939 s	170 489 s	178 804 s	185 952 s	4.0	5.8
IS	188	251 e	261	280 f	7.3	14.1
NO	2 445	:	3 037	:	:	11.5
JP	123 912	153 852	:	:	:	24.2
US	229 004	287 111	315 189	308 987	-2.0	10.5

NB: Annual growth rates — AGR — and annual average growth rates — AAGR — are calculated in current EUR.
ACC excludes MT.
(1) AGR: Annual growth rate between the two last available years.
(2) AAGR: Annual average growth rate between the last available year and 1999;
Exception to the reference period 1999-2002 — EL: 1997-99.

Sources: Eurostat; OECD.

In 2001, more than half of the R&D expenditure in absolute terms in the Candidate Countries was carried out in Poland and the Czech Republic – Table 2.2. The majority of Candidate Countries have annual average growth rates of R&D expenditure in EUR above 8% between 1999 and 2001. The expenditure increased by more than 24% from 2000 to 2001 in Hungary, Estonia and Lithuania.

However, the situation changes when calculated in real terms – constant 1995 PPS – Table 2.3. Portugal, Ireland, Denmark and Belgium, with annual real growth rates of 8.6%, 7.4%, 7.4% and 7.2% respectively, were the countries where R&D activity increased the most in 2001. The Netherlands, the United Kingdom and France were the only countries where negative rates were observed according to the most recent available data. Rates for the latter two countries and Germany, which belongs to the leading countries in terms of volume, were below not only those of Japan (3.8%) and the United States (2.3%) but also the EU average (1.9%). For several countries, the annual growth rate returned to smaller figures after several years of very high rates. This is notably the case for Iceland, Finland and the Netherlands, and to a lesser extent Germany. In the BES, three countries, Portugal, Denmark and Italy stand out with very high annual growth rates above 7%. In the GOV, the levels reached were on average higher and 4 countries increased their R&D expenditure by over 20%: Ireland, Luxembourg, Iceland and the United Kingdom. In the HES, Luxembourger annual growth rates reached 63%.

Table 2.3.

Annual real growth rates (1) of R&D expenditure in %, all sectors
EU-15, Iceland, Norway, Japan and the United States
1999 to 2002 (2)

	All sectors				Business enterprise				Government				Higher education			
	1999	2000	2001	2002	1999	2000	2001	2002	1999	2000	2001	2002	1999	2000	2001	2002
EU-15	**5.4 s**	**5.4 s**	**3.6 s**	**1.9 s**	**7.6 s**	**5.7 s**	**4.4 s**	**1.4 s**	**-1.5 s**	**3.5 s**	**-0.7 s**	**3.5 s**	**3.8 s**	**6.2 s**	**3.7 s**	**2.7 s**
BE	6.5 er	7.8 ep	7.2 ep	:	7.5 e	9.6 ep	8.5 ep	3.0 ep	2.7 er	100.8 ep	4.0 ep	:	4.5 er	-10.5 ep	3.2 ep	:
DK	4.5 r	11.2 er	7.4	:	2.4 r	17.0 er	10.8	:	10.7 r	-8.2 r	1.2	:	5.6	9.0 r	0.9	:
DE	7.4 r	5.3 e	0.4 e	:	10.3 r	6.1 e	0.7 e	:	0.8 r	3.9	-0.7 e	:	1.7 r	2.9	0.0 ep	:
EL	19.4	:	:	:	26.0	:	:	:	14.9	:	:	:	18.1	:	:	:
ES	3.1	10.6 er	4.5 r	:	2.9	14.2 er	2.0 r	:	7.0	3.7 r	4.9 r	:	1.8	8.8 r	9.1 r	:
FR	3.7	3.8	4.5 p	-0.3 e	5.2	2.7	5.5 p	-1.8 e	0.9	-0.9	0.2 p	1.7 e	1.1	13.4	5.2 p	2.8 e
IE	6.3 e	5.6	7.4	:	7.8	4.1	2.4	:	-12.3	43.5	26.9 p	:	7.6 e	0.1	17.4	:
IT	-0.9 r	5.9	:	:	1.1 r	7.5	7.2	:	-5.9 r	4.3	9.8	:	-0.7 r	4.4	:	:
LU	:	:	:	:	:	:	:	:	:	:	23.6 r	:	:	:	62.8 r	:
NL	8.4 r	-0.8	:	:	12.8 r	0.4	-1.0 p	:	-4.1 l	-17.2	-1.6 p	:	4.7 r	10.3	:	:
AT	6.8 e	2.6 e	4.2	3.3	:	:	:	:	:	:	:	:	:	:	:	:
PT	:	:	8.6 e	:	:	:	28.6 e	:	:	:	-6.4	:	:	:	5.9	:
FI	16.0	10.8	0.8	4.2 f	17.7	15.3	1.1	3.6 f	11.7	2.9	-2.8	2.7 f	16.6	0.3	2.0	7.2 f
SE	5.5	:	:	:	4.0	:	:	:	2.5	:	:	:	11.3	:	:	:
UK	4.3	3.1 r	4.8 r	-1.1 f	7.4	-0.3 r	7.7 r	-5.3 r	-16.1	16.9 r	-16.3 r	22.8 f	7.2	6.8 r	8.2 r	1.6 f
EEA	**5.4 s**	**5.4 s**	**3.5 s**	**1.9 s**	**7.5 s**	**5.7 s**	**4.5 s**	**1.4 s**	**-1.5 s**	**3.4 s**	**-0.8 s**	**3.4 s**	**3.9 s**	**6.0 s**	**3.6 s**	**2.6 s**
IS	20.0	22.2 e	14.2	0.7 f	52.9	47.6 e	19.3	-2.2 f	-2.9	10.9 e	-16.3	23.0 f	0.5 r	-5.0 e	32.1	-13.9 f
NO	:	:	1.0	:	:	:	4.3	:	:	:	-1.7	:	:	:	-4.4	:
JP	0.6	3.8	:	:	0.0	4.2	:	:	7.3	4.2	:	:	0.6	1.6	:	:
US	6.1	6.4	4.0	2.3	6.5	7.0	2.8	0.4	1.6	-3.8	6.5	11.0	5.0	6.5	6.6	7.3

NB: Annual average growth rates — AAGR — are calculated in constant 1995 PPS.
(1) AGR: Annual growth rate between the two last available years.
(2) **Exception to the reference period 2000-2001**
PT and NO: 1999-2001.

Sources: Eurostat; OECD.

'Industry' is R&D's most important source of financing for the EU-15 countries, Iceland and Norway, although the share of 'Industry' compared to other sources of financing R&D varies per country. For 8 EEA countries it accounts for more than 50% of the total R&D expenditure. In contrast, the financing sources are more balanced in the Candidate Countries – Figure 2.4., all sectors. In Sweden and Finland more than 70% of R&D expenditure is financed by 'Industry', a proportion which is close to that of Japan (73%). In the Candidate Countries, the top figures for financing by 'Industry' are lower, as the Slovak Republic and Slovenia retained rates of approximately 55%. 'Government' financing, which is second to 'Industry' in terms of importance, comes at the top for six countries: Portugal, Greece, Austria, Poland, Hungary and Turkey. The remaining sources, 'Abroad' and 'Other national sources', are of minor importance for all countries, except for Greece, Austria, United Kingdom and Iceland where more than 15% of R&D expenditure is financed from 'Abroad'.

As far as the performance of R&D by the Business enterprise sector – BES – is concerned, it is interesting to look at the source of financing and the share of 'Industry' compared to other sources of financing. Figure 2.4. shows that for six EU-15 countries 90% or more of the R&D executed is being financed by 'Industry' (similar to the United States at 89%) while Finland is the only EU-15 country approaching Japan (98%). The United Kingdom, Austria and Romania, on the other hand, display 36% of R&D expenditure performed by 'Industry' being financed by non-industry sources, such as 'Government', 'Abroad' and 'Other national sources'. The share of government-financed R&D performed by the Business enterprise sector is particularly high in Romania (31%), Poland (30%) and Slovak Republic (21%).

Figure 2.4. **R&D expenditure by source of financing as a % of total, all sectors and BES EEA, Candidate Countries, Japan and the United States 2001 (1)**

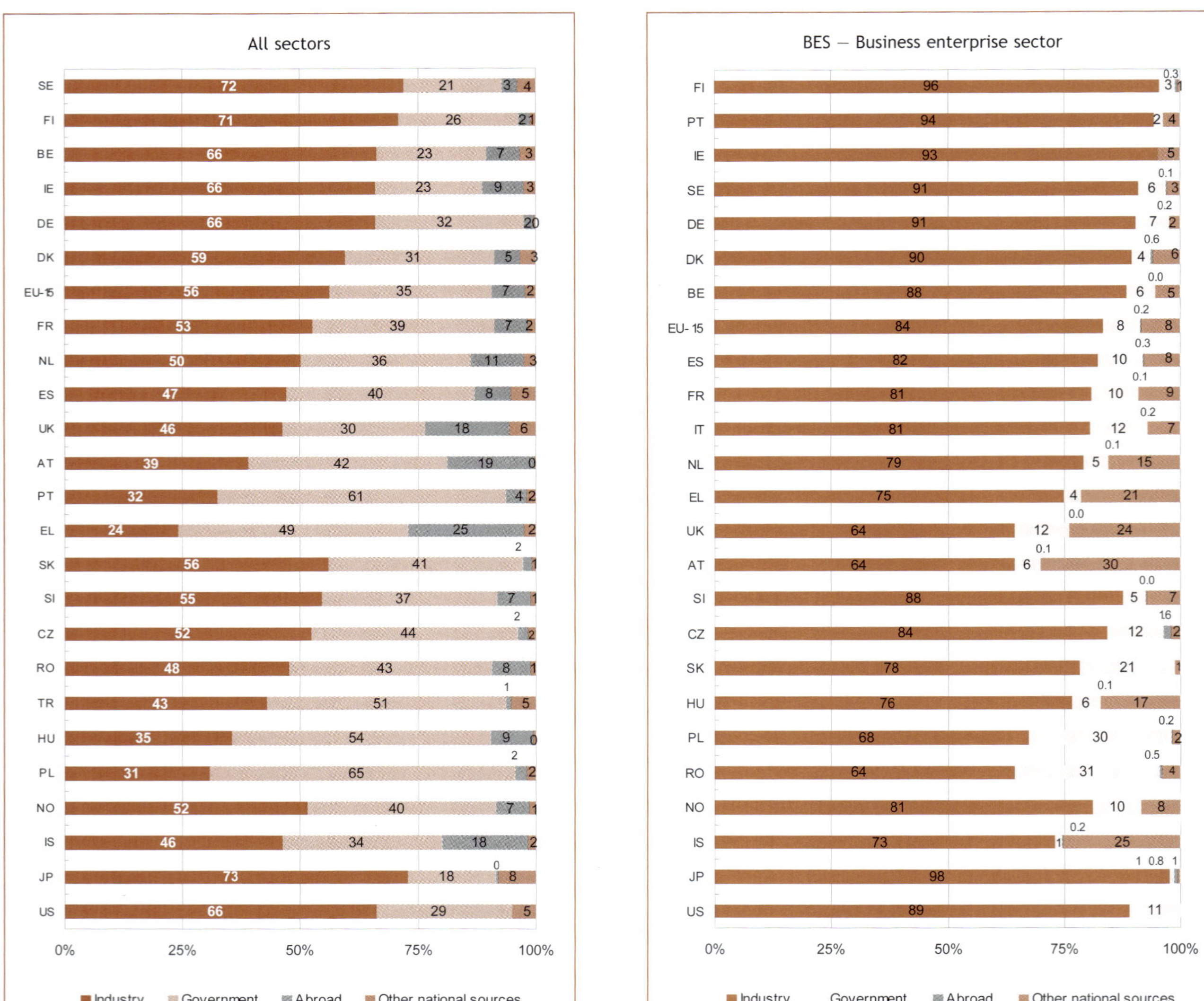

NB: Data not presented in the figures are not available.

(1) **Exceptions to reference year 2001**

All sectors
AT and US: 2002;
EU-15, FR, IE, NL and TR: 2000;
BE and DK: 1999.

Business enterprise sector — BES
IT and US: 2002;
EU-15, BE, FR, NL and TR: 2000;
DK and EL: 1999;
AT: 1998.

Sources: Eurostat; OECD.

2.4. R&D expenditure in the European regions

Regions with high R&D intensity and regional disparities

At the regional level, Braunschweig (6.21%) and Västsverige (5.27%) lead in terms of R&D intensity

According to Table 2.4., the top 10 R&D regions in the EU with the highest R&D expenditure as a percentage of GDP, i.e. R&D intensity, are mainly located in Germany, Sweden and Finland. The German region Braunschweig comes first with 6.21% which is three times the EU-15 average. Västsverige (SE) and Stuttgart (DE) follow with 5.27% and 4.82%, respectively. Only the British region of Eastern appears in this top ten with an R&D intensity of 3.56%, which is almost half of the figure for Braunschweig. The regions with high R&D intensity are also regions where the R&D activity is highly concentrated in terms of volume. For instance, the top four regions represent over 10% of the EU-15's total R&D expenditure. This figure reaches almost 20% when all the leading ten regions are considered.

Table 2.4. **Regions with a high level of R&D expenditure as a % of GDP, all sectors EU-15, Iceland and Norway 2001**

Regions	Country	As a % of GDP	Constant 1995 PPS	
			Mio	% of EU-15
EU-15		1.98	147 998	100.00
EEA		1.97	150 030	101.37
Braunschweig — 1999	DE	6.21	2 116	1.56
Västsverige	SE	5.27	1 958	1.32
Stuttgart — 1999	DE	4.82	4 807	3.55
Oberbayern — 1999	DE	4.72	5 578	4.12
Pohjois-Suomi	FI	4.36	464	0.31
Stockholm	SE	4.33	2 407	1.63
Tübingen — 1999	DE	4.22	1 563	1.15
Uusimaa (Suuralue)	FI	4.21	1 745	1.18
Berlin — 1999	DE	3.68	2 356	1.74
Eastern — 1999, NUTS 1	UK	3.56	3 745	2.76

Source: Eurostat.

R&D intensity at the national level shows divergent performances by country when the leading region of each country is taken into account – Figure 2.5. For all sectors together three main groups of countries may be pictured. At the top, Germany, Sweden and Finland stand out with an R&D intensity in their leading region above 4%. The second group includes countries for which the R&D intensity in their leading region is between the EU average of 1.99% and 4%. This group includes countries with high R&D expenditure in volume like the United Kingdom and France. Finally, the top region of five countries shows an R&D intensity rate below the EU-15 average, namely that of Italy, Luxembourg, Spain, Ireland, Portugal and Greece. Disparities exist not only among countries but also within regions of the same country. The gap between the leading region and the region at the bottom of the ranking is the largest in Germany where it reaches 5.8 percentage points whereas it is lowest in Portugal with 0.8 percentage points. With the exception of Northern Ireland in the United Kingdom, whose R&D expenditure almost reaches 1% of the GDP, R&D intensity in all the other lowest regions of Germany, Greece, Spain, France, Italy, the Netherlands, Austria, Portugal, Finland and Sweden, is less than 0.52%. – Figure 2.5.

Regional disparities also exist by institutional sector. The situation in the BES is similar to that described for the total of the sectors, the top region for eight countries remaining unchanged – Figure 2.6. In the GOV, the gaps between countries are less important with the exception of Flevoland (NL), which comes far ahead of a group of three regions belonging to the leading countries in terms of R&D expenditure: Dresden (DE), Languedoc-Roussillon (FR) and Lazio (IT). Groningen stands out in the HES with an R&D intensity of 1.75% far ahead of the other leading regions for which the R&D intensity is equal or below 1.1% – Figure 2.6.

Figure 2.5. **Regional disparities in R&D expenditure as a % of GDP, all sectors EU-15, Iceland and Norway 2001 (1, 2)**

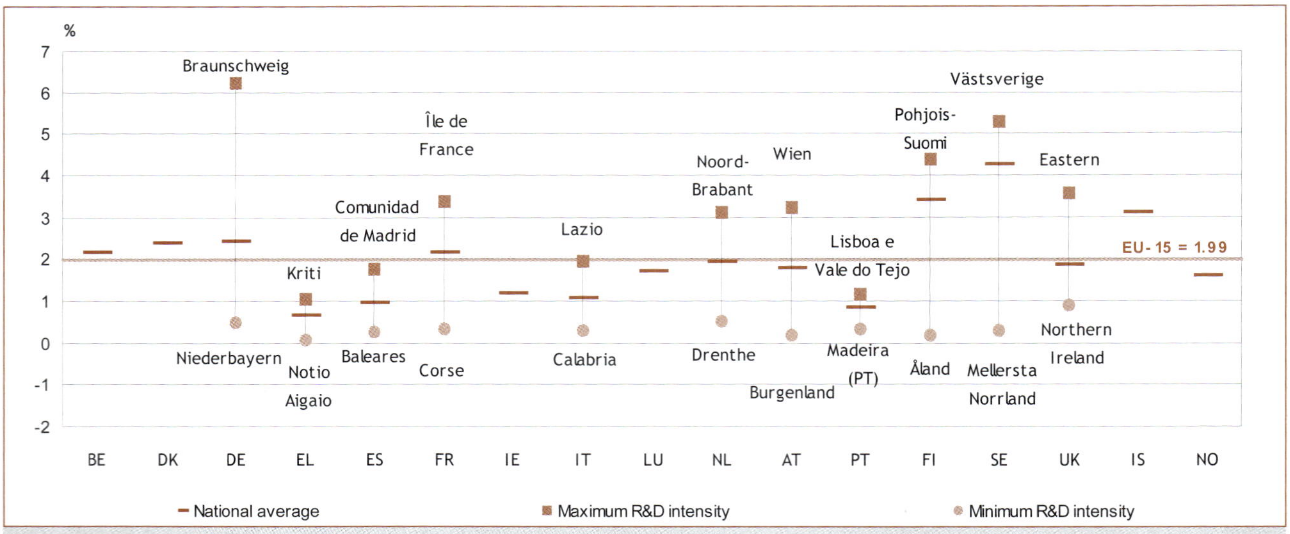

(1) **Exceptions to the reference year 2001**
EU-15 and IS: 2002;
FR, IT, LU and NL: 2000;
DE, EL and UK: 1999;
AT: 1998.

(2) **Exceptions to data at the NUTS 2 level**
IE and UK: classified at NUTS 1 level.

Source: Eurostat.

Figure 2.6. **Regional disparities in R&D expenditure as a % of GDP, BES, GOV and HES EU-15, Iceland and Norway 2001 (1, 2)**

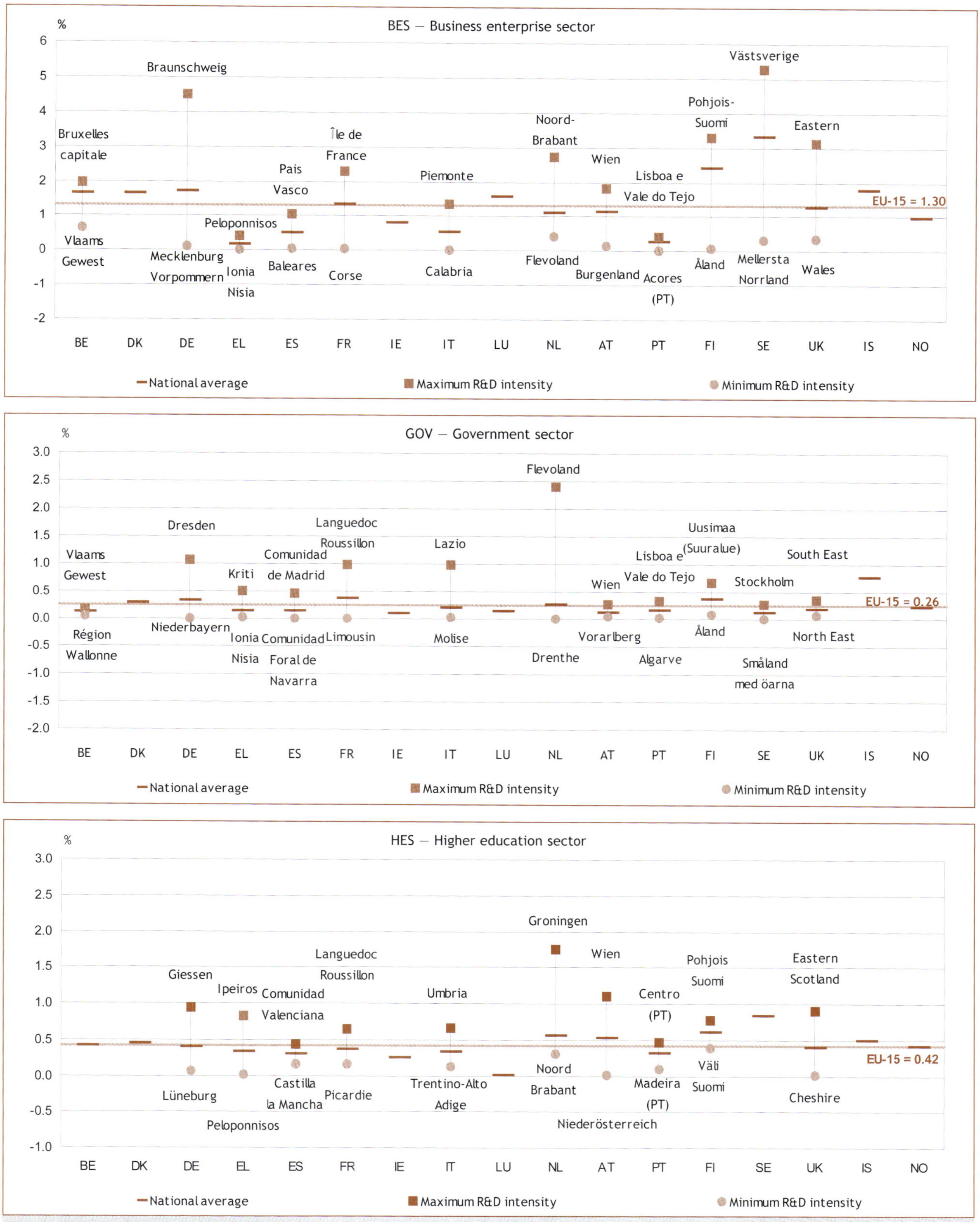

(1) **Exceptions to the reference year 2001**
BES — EU-15, BE and IS: 2002; FR, IT, LU and NL: 2000; DE and EL: 1999; AT: 1998;
GOV — EU-15 and IS: 2002; DE, FR, IT, and NL: 2000; EL and SE: 1999; AT: 1998;
HES — EU-15 and IS: 2002; DE, FR, IT, and NL: 2000; EL: 1999; AT: 1998.

(2) **Exceptions to data at the NUTS 2 level**
BES — BE, IE and UK: classified at NUTS 1 level;
GOV — BE and IE: classified at NUTS 1 level;
HES — IE: classified at NUTS 1 level.

Source: Eurostat.

Map 2.1.

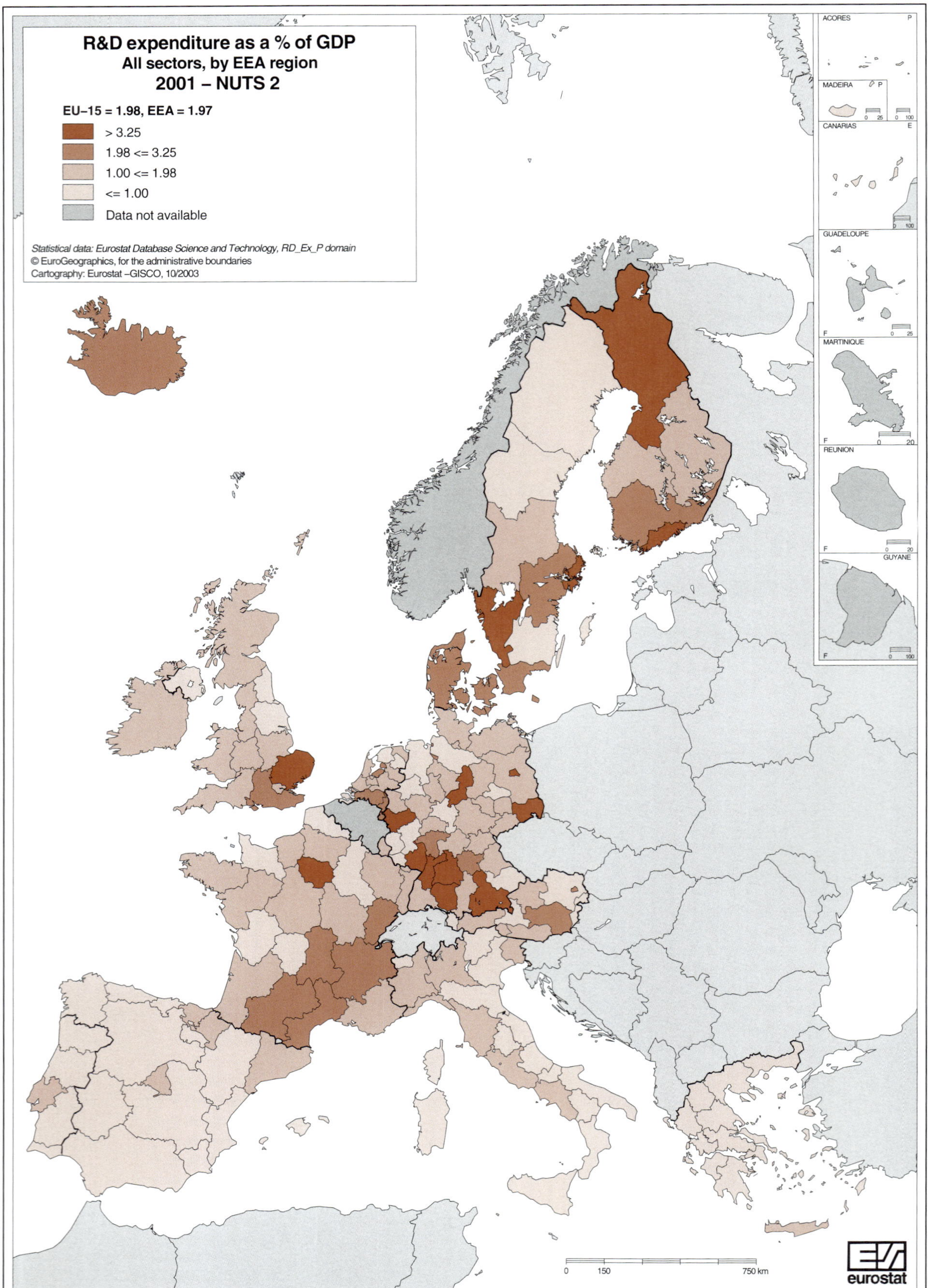

EU-15 = 1.98 and EEA = 1.97
refer to the EU-15 and the EEA averages, i.e. in 2001 and for all sectors, R&D expenditure in the EU and the EEA amounted to 1.98% and 1.97% of their GDP respectively.

Exceptions to the reference year 2001
EU-15 and IS: 2002; FR, IT, LU and NL: 2000;
DE, EL and UK: 1999; AT: 1998.

Exceptions to data at the NUTS 2 level
IE and UK: classified at NUTS 1 level.

Map 2.2.

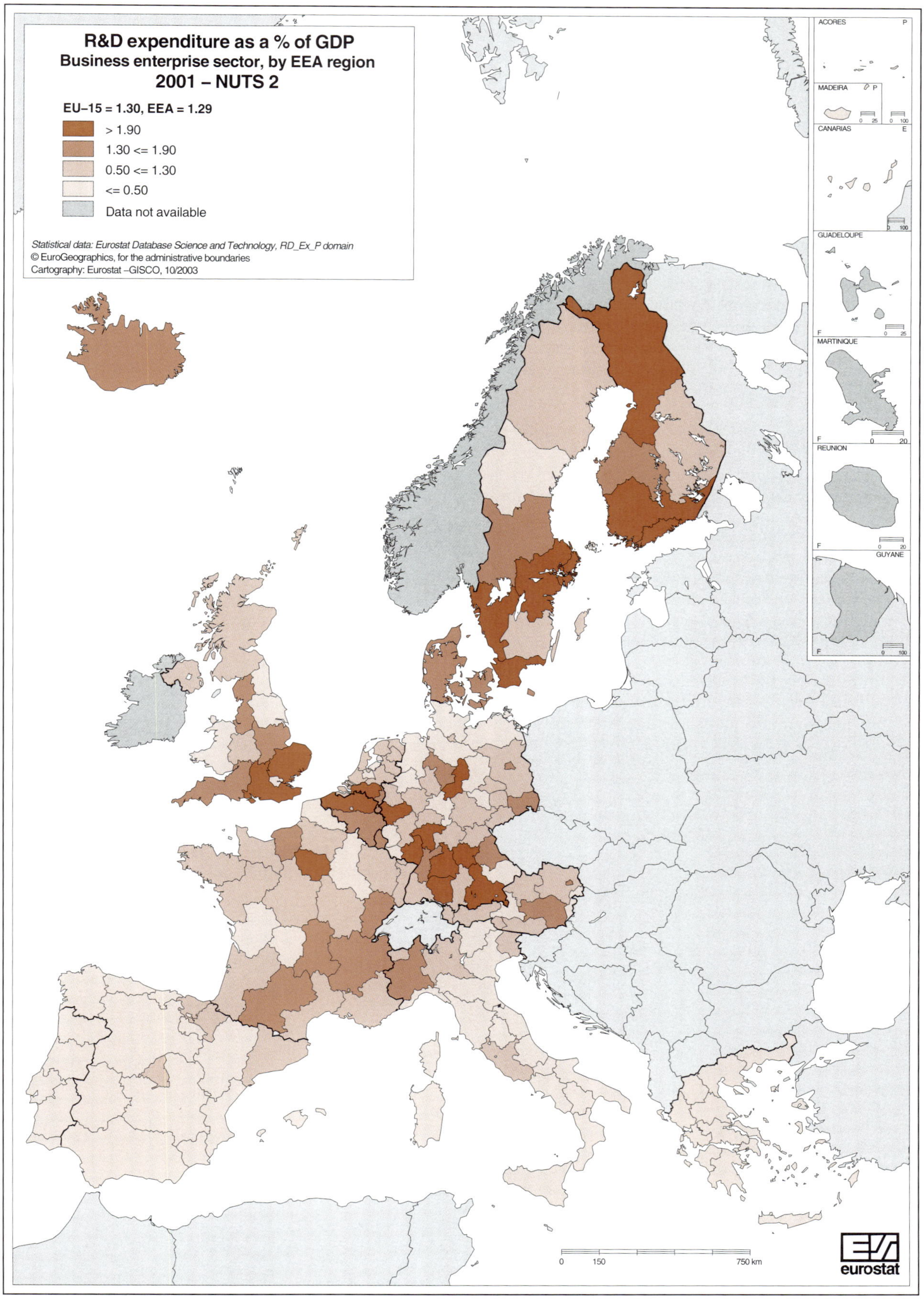

EU-15 = 1.30 and EEA = 1.29
refer to the EU-15 and the EEA averages, i.e. in 2001 and in the Business enterprise sector — BES, R&D expenditure in the EU and the EEA amounted to 1.30% and 1.29% of their GDP respectively.

Exceptions to the reference year 2001
EU-15, BE and IS: 2002; FR, IT, LU and NL: 2000; DE and EL: 1999; AT:1998.

Exceptions to data at the NUTS 2 level
BE, IE and UK: classified at NUTS 1 level.

Regions with high R&D expenditure in volume and regional disparities

R&D expenditure is concentrated in the leading regions, as the top 10 in 2001 accounted for 30% of the EU's total

In 2001, 30% of R&D expenditure in the EU was concentrated in ten regions when measured in constant 1995 PPS. Five of these regions were German, two were French the others were Italian, Danish and Swedish. Most of the R&D was carried out in Île de France (FR), as R&D expenditure in this region accounted for 8.1% of the total R&D expenditure in the EEA. Following Île de France were Oberbayern (DE, 4.1%) and Stuttgart (DE, 3.5%) – Figure 2.7.

Table 2.5. analyses the regional R&D activity within a country at a more detailed level, by showing the top three regions for each country in millions of EUR. It should be noticed that the number of regions varies from one country to another.

For all sectors, among the countries with very high level of R&D expenditure in volume, namely Germany, France, United Kingdom and Italy, Île de France has the highest regional concentration of R&D as it accounts for 44% of total R&D expenditure in France. In Germany the top region of Oberbayern represents only 14% of R&D expenditure, whereas in Italy, Lombardia has 22% of R&D expenditure. Very high levels of R&D concentration, around or superior to 45% for the leading region, were observed in Portugal – Lisboa e Vale do Tejo, Greece – Attiki, Austria – Wien, Finland – Uusimaa (Suuralue) and Norway – Oslo og Akershus.

The breakdown by institutional sector shows a different picture depending on the sector. The GOV is the sector where most of the R&D expenditure for a country is carried out predominantly by one region. It is also in the GOV, where the proportions are highest. More than 50% of R&D expenditure is concentrated in the leading region of six EU countries, the top value being reached by Portugal with Lisboa e Vale do Tejo, where 86% of the Portuguese R&D expenditure was concentrated in 2001.

The lowest concentration of R&D expenditure in the leading region by country was observed in the HES. Among EU regions, Wien (AT) retained the highest concentration of R&D expenditure (56% of the total in Austria). Compared to the other two sectors, the regional R&D expenditure breakdown for the top regions seems to be better balanced for each country. In general, the regional concentration in the BES is less prevalent than in the GOV but more prevalent than in the HES. Indeed, even if in some countries like Portugal or Greece more than 60% of the R&D expenditure is carried out by one region, the proportions spent in the other regions are equally substantial.

In most countries of the EU, both public and private R&D expenditure are generally concentrated in one region. This is the case for Greece with Attiki, Spain – Comunidad de Madrid, France – Île de France, Austria – Wien, Portugal – Lisboa e Vale do Tejo. Finland – Uusimaa (Suuralue) and Norway – Oslo og Akershus. The exceptions are the Netherlands, Germany and Italy where the leading region in the BES and the GOV is different depending on the sector.

Figure 2.7. **Percentage of R&D expenditure in the top 10 EEA regions in constant 1995 PPS, all sectors 2001**

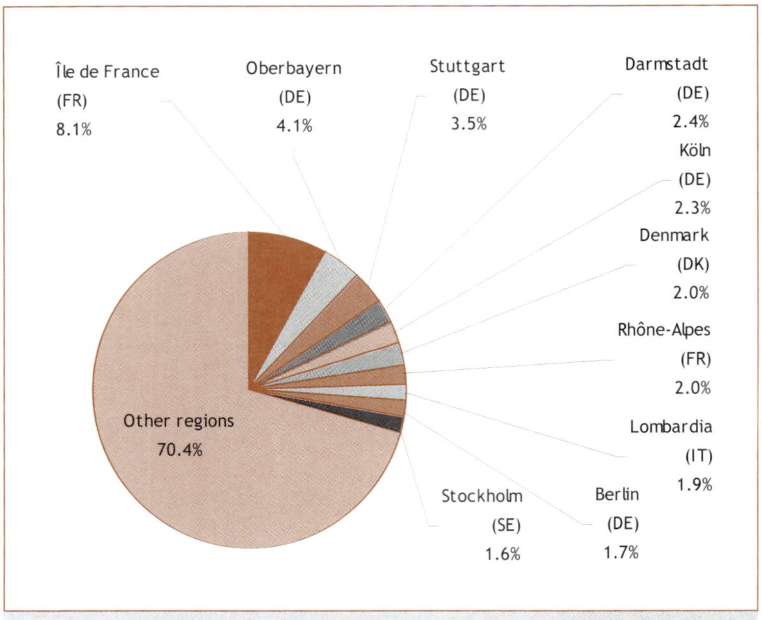

Source: Eurostat.

Table 2.5. **R&D expenditure in million EUR in the top 3 regions of each country, by institutional sector EU-15, Iceland and Norway 2001**

All sectors			Business enterprise sector			Government sector			Higher education sector		
Regions by country	Mio EUR	%	Regions by country	Mio EUR	%	Regions by country	Mio EUR	%	Regions by country	Mio EUR	%
EU-15	182 387		EU-15	119 000		EU-15	23 949		EU-15	38 197	
Belgium	5 515	100	Belgium	4 062	100	Belgium	331	100	Belgium	1 059	100
Denmark	4 265	100	Denmark	2 934	100	Denmark	503	100	Denmark	796	100
Germany – 1999	48 191	100	Germany – 1999	33 623	100	Germany – 2000	6 873	100	Germany – 2000	8 146	100
Oberbayern	6 548	14	Oberbayern	5 155	15	Köln	910	13	Oberbayern	698	9
Stuttgart	5 643	12	Stuttgart	5 104	15	Berlin	760	11	Köln	682	8
Darmstadt	3 868	8	Darmstadt	3 383	10	Karlsruhe	735	11	Berlin	610	7
Greece – 1999	795	100	Greece – 1999	226	100	Greece – 1999	173	100	Greece – 1999	394	100
Attiki	419	53	Attiki	144	63	Attiki	108	63	Attiki	163	42
Kentriki Makedonia	126	16	Peloponnisos	26	11	Kriti	31	18	Kentriki Makedonia	89	23
Kriti	64	8	Kentriki Makedonia	21	9	Kentriki Makedonia	15	9	Dytiki Ellada	46	12
Spain	6 227	100	Spain	3 261	100	Spain	989	100	Spain	1 925	100
Comunidad de Madrid	1 974	32	Comunidad de Madrid	1 096	34	Comunidad de Madrid	511	52	Comunidad de Madrid	344	18
Cataluna	1 334	21	Cataluna	891	27	Andalucia	119	12	Cataluna	331	17
Pais Vasco	561	9	Pais Vasco	434	13	Cataluna	104	11	Comunidad Valenciana	273	14
France – 2000	30 954	100	France – 2000	19 348	100	France – 2000	5 361	100	France – 2000	5 804	100
Île de France	13 474	44	Île de France	9 237	48	Île de France	1 978	37	Île de France	2 115	36
Rhône-Alpes	3 281	11	Rhône-Alpes	2 205	11	Midi-Pyrénées	468	9	Rhône-Alpes	692	12
Provence-Alpes-Côte d'Azur	1 807	6	Provence-Alpes-Côte d'Azur	988	5	Provence-Alpes-Côte d'Azur	422	8	Provence-Alpes-Côte d'Azur	398	7
Ireland	1 339	100	Ireland	917	100	Ireland	128	100	Ireland	294	100
Italy – 2000	12 460	100	Italy – 2000	6 239	100	Italy – 2000	2 356	100	Italy – 2000	3 865	100
Lombardia	2 793	22	Lombardia	2 066	33	Lazio	1 149	49	Lombardia	448	12
Lazio	2 309	19	Piemonte	1 364	22	Lombardia	279	12	Lazio	439	11
Piemonte	1 662	13	Lazio	721	12	Toscana	138	6	Toscana	429	11
Luxembourg – 2000	364	100	Luxembourg – 2000	337	100	Luxembourg – 2000	33	100	Luxembourg – 2000	2	100
Netherlands – 2000	7 813	100	Netherlands – 2000	4 457	100	Netherlands – 2000	1 078	100	Netherlands – 2000	2 278	100
Noord-Brabant	1 832	23	Noord-Brabant	1 610	36	Zuid-Holland	319	30	Zuid-Holland	585	26
Zuid-Holland	1 549	20	Noord-Holland	771	17	Noord-Holland	244	23	Noord-Holland	448	20
Noord-Holland	1 463	19	Zuid-Holland	645	14	Utrecht	194	18	Gelderland	315	14
Austria – 1998	3 377	100	Austria – 1998	2 146	100	Austria – 1998	218	100	Austria – 1998	1 003	100
Wien	1 639	49	Wien	934	43	Wien	136	63	Wien	562	56
Steiermark	596	18	Steiermark	361	17	Steiermark	22	10	Steiermark	214	21
Oberösterreich	392	12	Oberösterreich	332	15	Niederösterreich	14	6	Tirol	120	12
Portugal	1 038	100	Portugal	330	100	Portugal	216	100	Portugal	381	100
Lisboa e Vale do Tejo	625	60	Lisboa e Vale do Tejo	233	71	Lisboa e Vale do Tejo	185	86	Lisboa e Vale do Tejo	162	42
Norte	213	20	Norte	58	18	Norte	11	5	Norte	105	28
Centro (PT)	139	13	Centro (PT)	34	10	Centro (PT)	6	3	Centro (PT)	77	20
Finland	4 619	100	Finland	3 284	100	Finland	501	100	Finland	834	100
Uusimaa (Suuralue)	2 123	46	Uusimaa (Suuralue)	1 450	44	Uusimaa (Suuralue)	332	66	Uusimaa (Suuralue)	341	41
Etelä-Suomi	1 440	31	Etelä-Suomi	1 104	34	Etelä-Suomi	85	17	Etelä-Suomi	251	30
Pohjois-Suomi	565	12	Pohjois-Suomi	426	13	Pohjois-Suomi	39	8	Pohjois-Suomi	100	12
Sweden	10 459	100	Sweden	8 118	100	Sweden	289	100	Sweden	2 033	100
Stockholm	3 005	29	Stockholm	3 005	37	Stockholm	173	60			
Västsverige	2 445	23	Västsverige	2 445	30	Östra Mellansverige	69	24			
Sydsverige	1 017	10	Sydsverige	1 017	13	Övre Norrland	21	7			
United Kingdom	30 501	100	United Kingdom	20 393	100	United Kingdom	2 941	100	United Kingdom	6 489	100
									Inner London	1 510	23
									Berkshire, Bucks & Oxfords.	517	8
									Eastern Scotland	461	7
EEA	185 952		EEA	121 161			24 484		EEA	39 060	
Iceland – 2002	280	100	Iceland – 2002	160	100	Iceland – 2002	69	100	Iceland – 2002	45	100
Norway	3 037	100	Norway	1 814	100	Norway	444	100	Norway	780	100
Oslo og Akershus	1 352	45	Oslo og Akershus	777	43	Oslo og Akershus	250	56	Oslo og Akershus	325	42
Vestlandet	438	14	Soer-Oestlandet	288	16	Vestlandet	87	20	Vestlandet	161	21
Troendelag	421	14	Troendelag	239	13	Nord-Norge	33	7	Troendelag	157	20

Source: Eurostat.

PART 2

Human resources in R&D in Europe

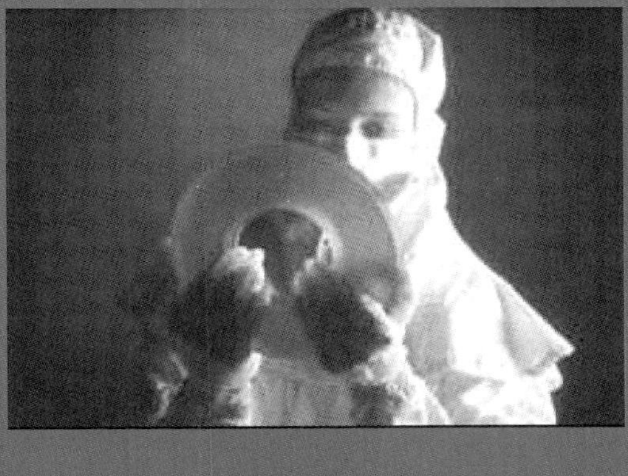

Chapter 3

3.1. Introduction

As seen in Section 2.1., R&D activities are often considered a catalyst for economic growth as they comprise creative work undertaken on a systematic basis in order to increase the stock of knowledge, including knowledge of man, culture and society, and the use of this stock of knowledge to devise new applications.

The number of R&D personnel is one of the two basic R&D input indicators together with R&D expenditure. It is based on the definitions given in the *Frascati Manual* (1), which is the methodological reference for surveys on research and experimental development. However, R&D personnel is an indicator that covers the measurement of the personnel that participate to a nation's S&T development only partially. For international S&T indicators, another approach that covers a much broader population than the concept of R&D personnel as defined by the *Frascati Manual* exists. This is the concept of Human Resources in Science and Technology, which are analysed in Chapter 4 and Chapter 6 of this publication.

Being a key element of knowledge, S&T dissemination and development, R&D personnel has become an indicator increasingly appreciated by policy makers. As a consequence, the list of indicators collected and disseminated by the European Commission has also augmented in recent years. In this sense, from total R&D personnel by institutional sector, the information is now detailed by occupation, gender and field of science including cross-indicator breakdowns. Not all of these data series appear systematically in this publication, as the chapter focuses on the main trends shown by the key R&D personnel indicators, including one of the derived indicators which is R&D personnel as a proportion of the labour force. However, the complete series are available on CD-ROM (2) and on Eurostat's reference database *NewCronos*.

In view of the enlargement of the EU planned for 2004, a common analysis of R&D personnel in EU and Candidate Countries, including an aggregate for Acceding Countries, is also presented in this chapter. In order to facilitate the reading, in the text the term Candidate Countries refers to both ten Acceding countries and three Candidate Countries.

R&D personnel data measure the human resources going directly into R&D activities (3). As recommended by the *Frascati Manual*, R&D personnel data are provided in both full-time equivalent – FTE– and head count – HC. Data are available at both national and regional levels on the basis of the *Nomenclature of territorial units for statistics* – NUTS (4). The methodological guidelines in the field of regional statistics on R&D are given in the *Regional Manual* (5).

R&D personnel data are broken down by institutional sector, i.e. by sector engaged in R&D. In this publication, five sectors are used to calculate indicators of R&D activity:

- the Business enterprise sector — BES,
- the Government sector — GOV,
- the Higher education sector — HES,
- the Private non-profit sector — PNP and
- all sectors, which corresponds to the sum of the four previous sectors.

However, given the minor role played by the PNP sector in all countries, except in Portugal, it has not been systematically included in all the analyses in this chapter.

This chapter is divided into three parts.

- Firstly an international overview of the composition and evolution of R&D personnel is given by comparing data for the EU with that of its main competitors, Japan and the United States.
- Then the chapter focuses on the performance of the Member States of the EU, Candidate Countries, Iceland and Norway.
- Finally, a regional perspective of the distribution and evolution of R&D personnel in the EEA is given.

The analysis in this chapter covers the period from 1993 to 2002.

(1) ***Standard method for surveys on R&D and experimental development — Frascati Manual,*** OECD, 2002.

(2) ***Statistics on science and technology, 2003 Edition, Data 1980-2002,*** Eurostat, 2003.

(3) ***Standard method for surveys on R&D and experimental development — Frascati Manual,*** OECD, 2002, Section 5.1.

(4) ***Nomenclature of territorial units for statistics — NUTS,*** Eurostat, 1999.

(5) ***The regional dimension of R&D statistics and innovation — Regional Manual,*** Eurostat, 1996.

3.2. R&D personnel in the EU, Japan and the United States

R&D personnel

R&D personnel in the EU is on an upward trend. The number of researchers in the EU is growing, but still behind the United States

In 2002, 2.46 million people, expressed in head count – HC – were employed in the field of R&D in EU-15. When measured in full-time equivalent – FTE, the EU's R&D personnel contingent amounted to 1.83 million, which represented an increase of 1.6% compared to the previous year. Over the period 1998 to 2002, R&D personnel in the EU increased by 9.7%. The increase of R&D personnel in the EU during the 1998-2002 period was above that of Japan from 1997-2000, where only a raise of 0.3% was observed. Indeed, for the second consequent year, Japan recorded a drop in its R&D personnel: whilst in 1999 0.92 million people measured in FTE were employed in research, they amounted to 0.90 million in 2000 – Figure 3.1.

In the long run, the EU's R&D personnel contingent in FTE increased by 255 705 people between 1993 and 2002, which corresponds to a growth of 16%. The trend is the opposite in Japan, where the total of R&D personnel employed lowered by 50 608 people or 5.3% between 1993 and 2000.

When the distribution of R&D personnel by institutional sector is looked at, in 2002 55% of the R&D personnel In the EU were employed in the BES, 30% in the HES and 14% in the GOV. The proportion accounted for by each institutional sector varies slightly in Japan, where in 2000 the BES employed almost 2/3 of the total R&D personnel. The public sector therefore accounted for a lower proportion of R&D personnel: 7% in the GOV and 25% in the HES.

The distribution of R&D personnel by the institutional sector was stable over the last years for both EU-15 and Japan. For the period 1997-2002, only slight changes were observed in the EU, an increase of 1% in the HES which counter balances a corresponding decrease in the GOV. For Japan, the share of the BES decreased by 1% in 2000, reflecting the decrease of the R&D personnel that was observed in Figure 3.1. In parallel, the percentage of R&D personnel employed in the GOV increased by 1% – Figure 3.3.

Figure 3.1.

R&D personnel in FTE and HC, all sectors
EU-15 and Japan (1)
1993 to 2002

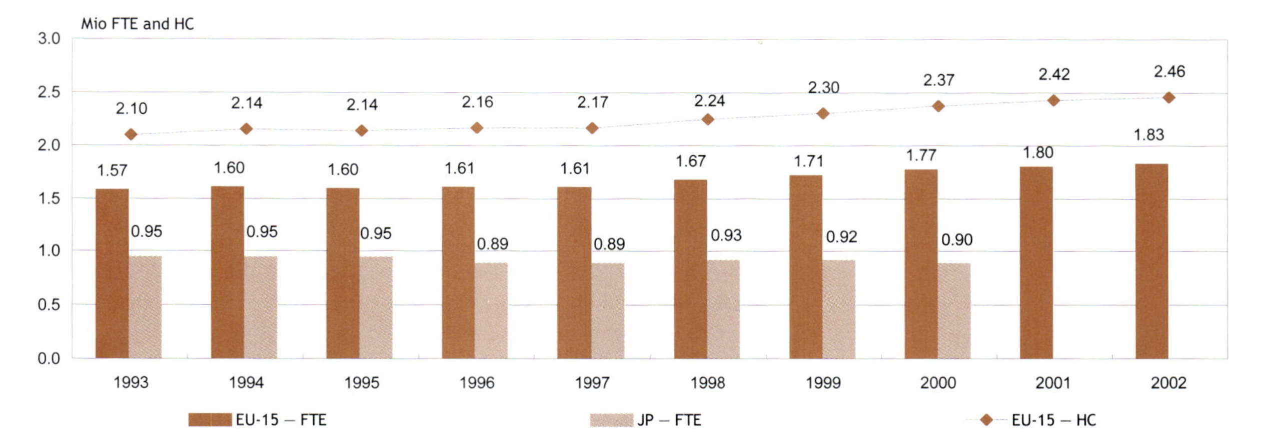

(1) EU-15: Eurostat estimates.
JP — 1993-95: overestimated or based on overestimated data; 1996: break in series.

Sources: Eurostat, OECD.

Figure 3.2.

Distribution of R&D personnel in FTE, by institutional sector
EU-15 (1)
1997 to 2002

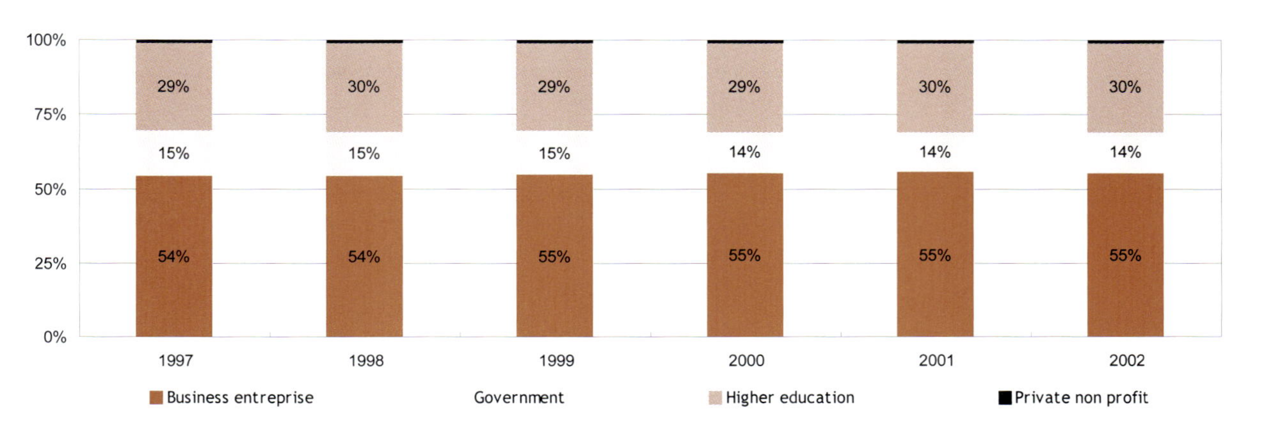

(1) EU-15: Eurostat estimates.

Source: Eurostat.

Figure 3.3.

Distribution of R&D personnel in FTE, by institutional sector
Japan
1997 to 2000

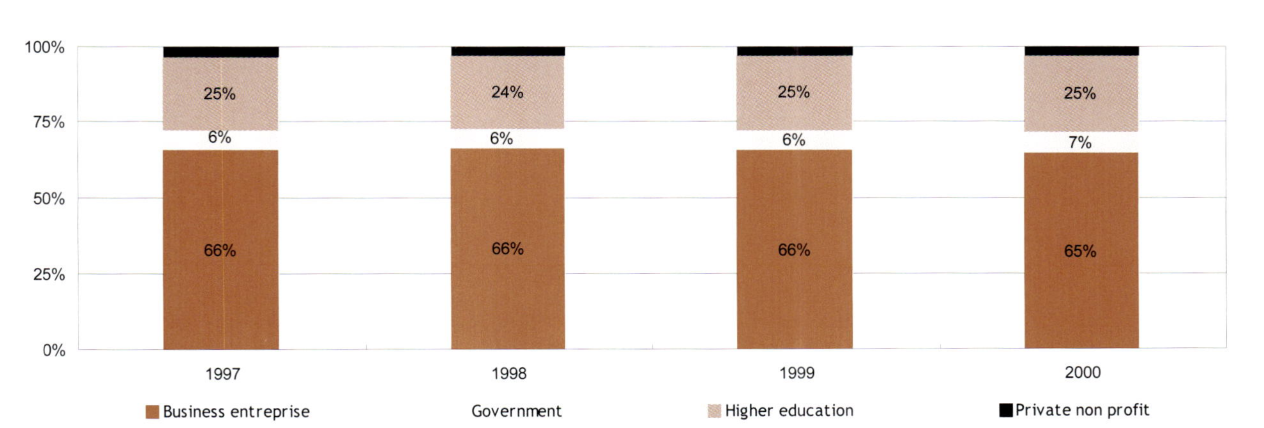

Source: OECD.

Researchers

The EU has the most researchers in the Government and Higher education sectors, whereas the United States leads in the Business enterprise sector

According to the latest available data on researchers measured in full-time equivalent – FTE, the United States employed the highest number of researchers in all sectors (1 114 100 people in 1997), compared to the EU (1 001 209 in 2002) and Japan (647 572 in 2000)– Figure 3.4. The number of researchers increased with regard to the previous year for both the United States and the EU by 6% and 2%, respectively. However, in Japan their number decreased by 1.7%.

Since 1995, the number of researchers in the United States showed a strong upward trend, which was also observed in the EU two years later. Indeed, between 1995 and 1997 the annual average growth rate recorded by the United States was 6.2% against the rate of 1.2% retained during the 1993-95 period. Similarly, the annual average growth rate of R&D personnel in the EU was 2% before 1997 and increased to 3.3% since then. With an annual average growth rate of 1.2% between 1997 and 2000, the trend for Japan remained positive but below that of the United States and the EU.

Differences across the three blocks are particularly noticeable when data are looked at broken down by institutional sector.

In the BES, the number of researchers in the United States is twice as high as in the EU or in Japan. On the contrary, the EU has the most researchers in the public sector. In the HES for instance, at 348 541, the number of researchers in the EU is 1.7 times higher than in the United States. The situation is similar in the GOV where 129 164 researchers are employed in EU against 47 700 in the United States and 31 228 in Japan.

In terms of trends by institutional sector, Japan's number of researchers remained relatively stable over the last decade in all of the sectors. The United States was the most dynamic country in the BES, as registered by an annual average growth rate of 5.6% between 1995 and 2000. In the HES, the EU appears as most dynamic, with an annual average growth rate of 1.5% during the 1997-2002 period.

Figure 3.4.

**Researchers in FTE, by institutional sector
EU-15, Japan and the United States (1)
1993 to 2002**

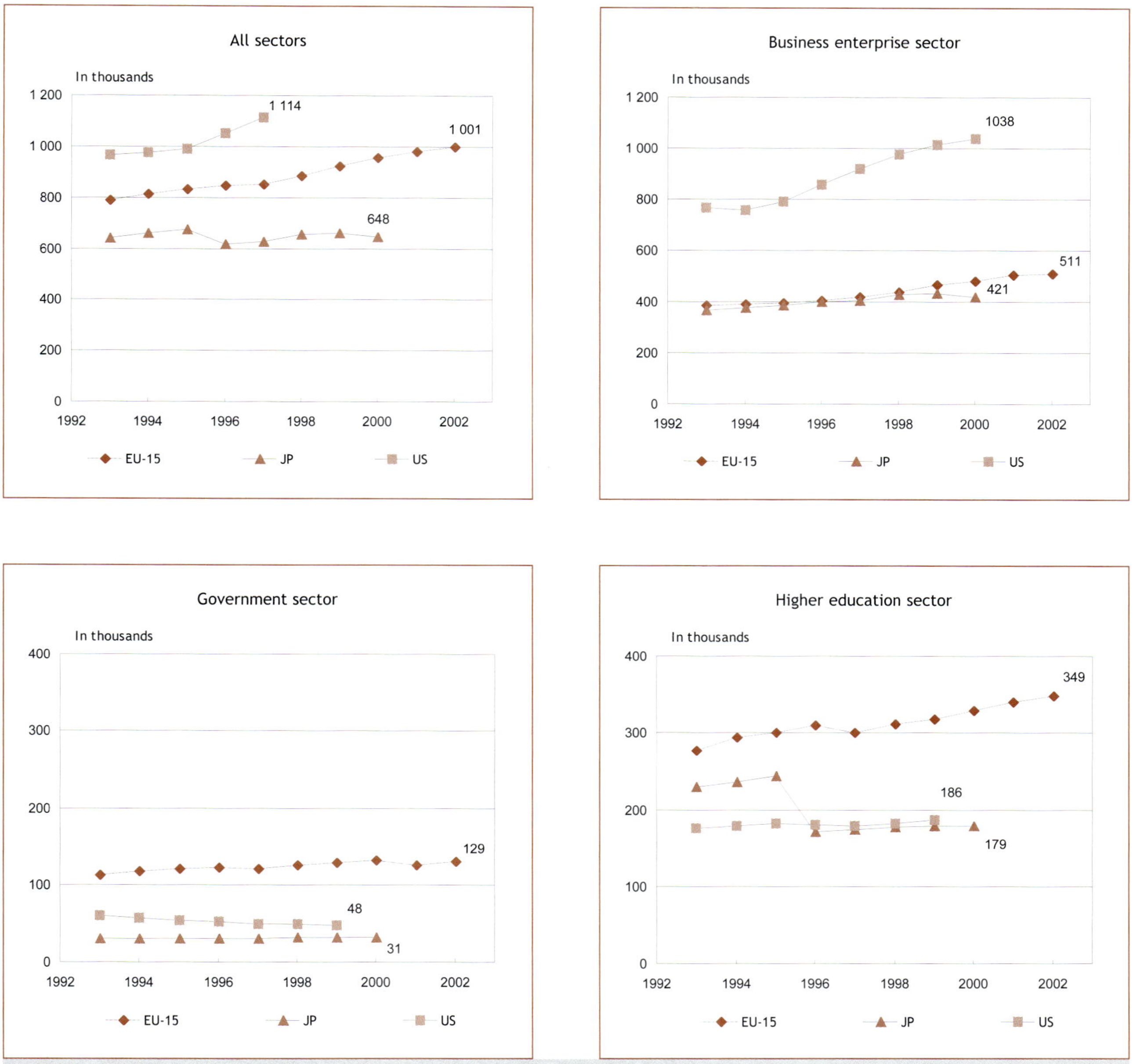

(1) **EU-15**
Eurostat estimates.
JP
All sectors and HES 1996: break in series;
BES and HES 1993-95: overestimated or based on overestimated data.
US
GOV 1993-99: Federal or central government only, defence excluded all or mostly.

Sources: Eurostat, OECD.

3.3. R&D personnel in Europe

R&D personnel as a % of labour force

With a percentage of employment in R&D that is almost double the EU average, Finland is leading

In 2002, 1.39% of the labour force in EU-15 and 0.84% in the Acceding Countries – ACC – worked in R&D. The EU gained 0.06 percentage points compared to 1999 whereas the ACC registered a more moderate increase of 0.02 percentage points.

Among the EU countries, the top three countries in the ranking did not change between 1999 and 2001: Finland leads with 2.60%, ahead of Sweden (2.43%) and Denmark (2.11%). For both reference years, the gap between the leading country and the EU average remained stable (1.2%). However, at the bottom of the ranking, only 2 EU countries, Ireland and Portugal, still have less than 1% of their labour force employed in R&D, against 4 countries in 1999. Three countries stand out with particularly high increases on the proportion of R&D personnel between 1999 and 2001: in Greece the share of R&D personnel in the labour force raised by 0.5%, whereas in both Denmark and Belgium it increased by 0.2%. R&D personnel as a percentage of the labour force figures for the top Candidate Countries was below the EU average in 1999 and also in 2001. The Candidate Countries with the highest ratios were Slovenia and Hungary, with 1.36% and 1.11%, respectively. The top 5 countries in the ranking among Acceding Countries remained unchanged over this period. In general, the share of R&D personnel in the labour force increased moderately, with Hungary and Slovenia recording the highest growth of 0.08% – Figure 3.5.

Figure 3.5.

**R&D personnel as a % of labour force, all sectors
EU-15, Candidate Countries, Iceland and Norway
1999 and 2001 (1, 2)**

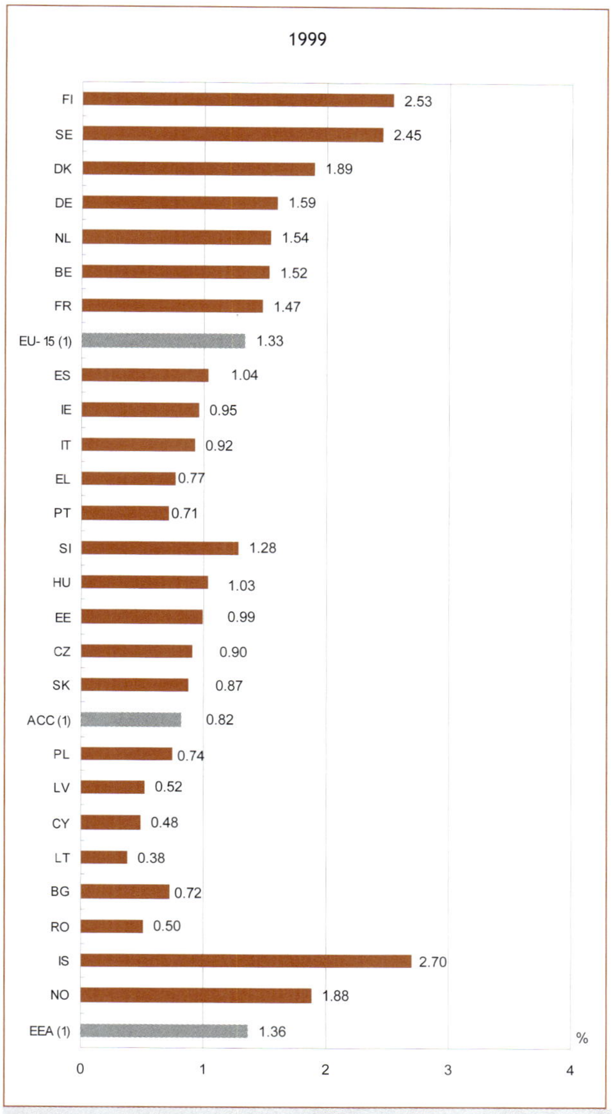

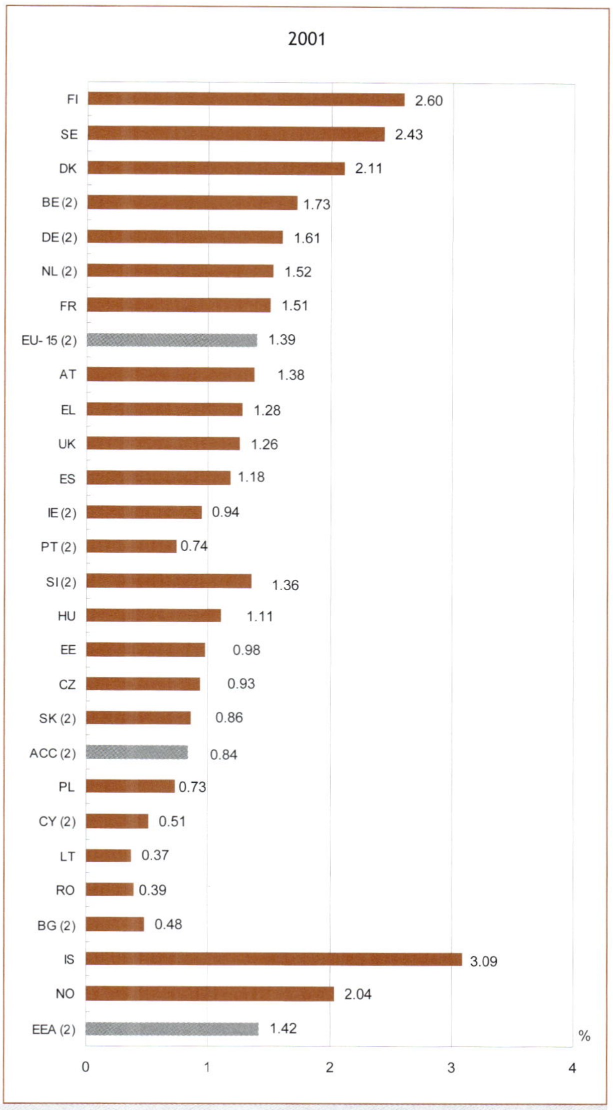

NB: Calculations have been made in head count — HC;
ACC excludes MT.

(1) **1999 data**
EU-15 and EEA: Eurostat estimates;
ACC: estimated values.
Exceptions to the reference year 1999
EL: 1997;
FR: 1996.

(2) **2001 data**
EU-15 and EEA: Eurostat estimates;
ACC: estimated values;
PT, CY, SI, SK and BG: estimated values;
BE, DE, IE and NL: Eurostat estimates.
Exceptions to the reference year 2001
EU-15 and EEA: 2002;
EL: 1999;
FR and AT: 1998;
UK: 1993;
All Candidate Countries: 2000 data except for LV and LT: 2001.

Source: Eurostat.

By institutional sector, the Business enterprise sector – BES – is the sector that employs the highest proportion of the labour force in R&D activities within the EU – Table 3.1. In 2002, as in 2001, the proportion in the BES remained stable at 0.67%. In the HES, this share in 2002 equalled to 0.54%, gaining 0.01% compared to the previous year. In the Government sector – GOV – in turn, only 0.17% of the labour force was R&D personnel. The share of R&D personnel by sector is rather different for the Acceding Countries, where the Higher education sector – HES – comes ahead with 0.46%. The BES employed 0.20% of its labour force in R&D whilst the rate registered on the GOV was 0.18%.

At the national level, the pictures shown by the EU countries and the Acceding Countries are rather different. In the EU countries the BES comes systematically ahead of the other sectors for all countries as opposed to the Acceding countries where the proportion of the R&D personnel varies more among the sectors.

Finland, Iceland, Denmark and Sweden are the top four EEA countries in the BES with percentages of R&D personnel superior to 1.1%, which represent the double of the figure reached by Slovenia, the leading Acceding Country for the BES. Except in Denmark where R&D personnel as a % of labour force in the BES increased by 0.08%, figures in general showed only small changes compared to the previous year.

In the HES, the figures for Sweden (1.14%), Greece and Iceland (both 0.91%), Norway (0.89%) and Finland (0.80%) are well above the EU average. The difference with the Candidate Countries is smaller than in the BES, with Estonia and Hungary displaying figures above the EU average. Both these countries belong to a group including Latvia, Lithuania, Poland and the Slovak Republic where most of the R&D personnel were employed in the HES. Finally, EU and Acceding Countries are on a similar level in the Government sector – GOV – as shown by their respective averages. Among all the countries, Iceland stands out in the GOV with a rate of 0.81%.

Table 3.1.

R&D personnel as a % of labour force, by institutional sector EU-15, Candidate Countries, Iceland and Norway 2000 to 2002 (1)

	All sectors			Business enterprise			Government			Higher education		
	2000	2001	2002	2000	2001	2002	2000	2001	2002	2000	2001	2002
EU-15	1.37 s	1.39 s	1.39 s	0.65 s	0.67 s	0.67 s	0.18 s	0.17 s	0.17 s	0.52 s	0.53 s	0.54 s
BE	1.60 s	1.73 s	:	0.88 s	0.97 s	0.97 s	0.11 s	0.11 s	:	0.60 s	0.63 s	:
DK	1.96 e	2.11	:	1.12 e	1.20	:	0.32 r	0.35	:	0.51 r	0.54	:
DE	1.61 s	1.61 s	:	0.92 s	0.92 s	:	0.24 s	0.24 s	:	0.45 s	0.45 s	:
EL	1.28	:	:	0.19	:	:	0.18	:	:	0.91	:	:
ES	:	1.18	:	:	0.31	:	0.17	0.18	:	0.64	0.68	:
FR	1.51	:	:	0.73	:	:	0.20	:	:	0.54	:	:
IE	0.94 s	:	:	0.58 s	0.59 s	:	0.10 s	0.11 s	:	0.26 s	:	:
IT	0.92	:	:	0.32	:	:	0.18	:	:	0.45	:	:
LU	:	:	:	:	:	:	0.18	0.20	:	0.02	0.03	:
NL	1.52 s	:	:	0.77	:	:	0.18	:	:	0.55 s	:	:
AT	1.38	:	:	0.65	:	:	0.15	:	:	0.57	:	:
PT	:	0.74 e	:	:	0.13 e	:	:	0.16	:	:	0.36	:
FI	2.58	2.60	:	1.43	1.42	:	0.38 i	0.38 i	:	0.77	0.80	:
SE	:	2.43	:	:	1.17	:	:	0.12	:	:	1.14	:
UK	1.26	:	:	0.56 s	0.59 s	0.56 s	0.11 s	0.08 s	0.09 s	0.40	:	:
ACC	0.84 f	:	:	0.20 f	:	:	0.18 f	:	:	0.46 f	:	:
CZ	0.93	:	:	0.39	:	:	0.23	:	:	0.30	:	:
EE	0.98	:	:	0.14	:	:	0.17	:	:	0.67	:	:
CY	0.51 f	:	:	0.14 f	:	:	0.20 f	:	:	0.12 f	:	:
LV	0.69	0.70	:	0.15	0.14	:	0.12	0.13	:	0.42	0.43	:
LT	0.36	0.37	:	0.02	0.02	:	0.12	0.12	:	0.22	0.23	:
HU	1.11	:	:	0.20	:	:	0.28	:	:	0.64	:	:
MT	:	:	:	:	:	:	:	:	:	:	:	:
PL	0.73	:	:	0.14	:	:	0.13	:	:	0.46	:	:
SI	1.36 f	:	:	0.54 f	:	:	0.35 f	:	:	0.46 f	:	:
SK	0.86 f	:	:	0.25 f	:	:	0.18 f	:	:	0.43 f	:	:
BG	0.48 f	:	:	0.06 f	:	:	0.32 f	:	:	0.09 f	:	:
RO	0.39	:	:	0.24	:	:	0.08	:	:	0.06	:	:
TR	:	:	:	:	:	:	:	:	:	:	:	:
EEA	1.39 s	1.42 s	1.42 s	0.65 s	0.67 s	0.67 s	0.18 s	0.18 s	0.18 s	0.54 s	0.55 s	0.56 s
IS	2.92 e	3.26	3.09 f	1.03 e	1.22	1.12 f	0.81 e	0.84	0.81 f	0.92 e	0.95	0.91 f
NO	:	2.04	:	:	0.87	:	:	0.27	:	:	0.89	:

NB: Calculations have been made in head count — HC.
ACC excludes MT.

(1) **Exceptions to the reference year 2000**
IT — all sectors and HES — and EL: 1999;
FR and AT: 1998;
UK — all sectors and HES: 1993.

Source: Eurostat.

R&D personnel in full-time equivalent – FTE

In 2002, R&D personnel in FTE in the EU increased by 1.6% compared to 2001

In 2001, Germany and France employed almost half of the EU's R&D personnel measured in full-time equivalent – FTE, as their R&D personnel contingent amounted to 487 378 and 333 517 people, respectively – Table 3.2. The United Kingdom came next with 277 500 – in 1993.

By institutional sector, if the proportion of R&D personnel in EU-15 accounted for by Germany for all sectors was 27% in 2001, this percentage reached 31% in the BES. Its weight was proportionally the least important in the HES where only 19% of the EU's R&D personnel was employed in Germany. Nevertheless, as seen in Table 3.1., when figures are observed as a proportion of the labour force, larger countries remain relatively close to the EU average. For all sectors for instance, Germany ranks at a 5th position and France is 7th out of the 15 Member States – see Table 3.1.

In absolute terms, however, Germany, France and the United Kingdom are leading in all institutional sectors, except in the Government sector – GOV, where Italy lays in third position ahead of the United Kingdom.

In the Candidate Countries, Poland employs most of R&D personnel – 78 925 persons, ahead of Romania – 32 639 and the Czech Republic – 26 107. However, in volume, the levels of R&D personnel for these leading countries remain in general far below those of most EU countries.

The Higher education sector – HES – is the most important sector for 7 Candidate Countries out of 12, as data for Malta are not available. It should be noted that for Estonia and Turkey, the HES accounts for more than 60% of their total R&D personnel. Only in Romania, the Czech Republic and Slovenia, the Business enterprise sector – BES – recorded the largest contingent of R&D personnel in FTE. The two Candidate Countries where the GOV employs most R&D personnel are Bulgaria and Cyprus, with more than half of the total number of R&D personnel employed in this sector.

Table 3.2.

R&D personnel in thousands of FTE, by institutional sector, EU-15, Candidate Countries, Iceland and Norway 2000 to 2002 (1)

	All sectors			Business enterprise			Government			Higher education		
	2000	2001	2002	2000	2001	2002	2000	2001	2002	2000	2001	2002
EU-15	1 765.8 s	1 801.2 s	1 829.6 s	971.9 s	999.5 s	1 014.2 s	252.9 s	246.6 s	249.1 s	520.0 s	535.0 s	545.6 s
BE	52.8 ep	55.9 ep	:	33.0 ep	35.5 ep	35.9 f	3.5 ep	3.7 ep	:	15.9 ep	16.2 ep	:
DK	37.7 e	39.9	:	23.7 e	25.8	:	5.7 r	5.5	:	8.0 r	8.3	:
DE	484.7 e	487.4 e	:	312.5 e	314.3 e	:	71.5	71.1 e	:	100.8	101.9 e	:
EL	26.4	:	:	4.6	:	:	4.4	:	:	17.3	:	:
ES	120.6 e	125.8	:	47.1 e	46.5	:	22.4 r	23.5	:	49.5 r	54.6	:
FR	326.4 r	333.5	:	181.0 r	185.5	:	49.8 r	49.4	:	89.0 r	91.9	:
IE	12.8	:	:	8.7	9.1	:	1.4	1.6 p	:	2.6	:	:
IT	150.1	:	:	64.0	:	:	31.2	:	:	54.8	:	:
LU	3.7	:	:	3.3	:	:	0.3	0.3	:	0.0	0.0	:
NL	88.5	:	:	47.5	60.1 p	:	14.2	14.0 p	:	26.7	:	:
AT	31.3	:	:	20.4	:	:	2.1	:	:	8.7	:	:
PT	:	23.0 e	:	:	3.9 e	:	:	6.0	:	:	10.2	:
FI	52.6	53.4	:	29.4	30.1	:	7.8 i	7.7 i	:	15.5	15.6	:
SE	:	72.1	:	:	49.4	:	:	2.8	:	:	19.8	:
UK	277.5	:	:	145.5 r	151.8	146.0 f	29.7 r	23.5 r	26.2 f	65.5	:	:
ACC	172.1	:	:	48.4	:	:	48.4	:	:	74.9	:	:
CZ	24.2	26.1	:	11.5	12.0	:	7.1	7.8	:	5.3	6.0	:
EE	3.7	3.7	:	0.4	0.6	:	0.9	0.7	:	2.3	2.3	:
CY	0.7	0.7	:	0.1	0.1	:	0.3	0.4	:	0.1	0.1	:
LV	5.4	5.5	:	1.4	1.4	:	1.2	1.1	:	2.9	3.0	:
LT	11.8	11.9	:	0.6	0.6	:	5.0	4.7	:	6.2	6.6	:
HU	23.5	22.9	:	6.5	6.8	:	8.2	7.8	:	8.9	8.4	:
MT	:	:	:	:	:	:	:	:	:	:	:	:
PL	78.9	:	:	18.6	:	:	18.8	:	:	41.5	:	:
SI	8.6	8.6	:	4.1	4.3	:	2.6	2.4	:	1.7	1.8	:
SK	15.2	14.4	:	5.2	4.8	:	4.2	4.0	:	5.9	5.7	:
BG	15.3	14.9	:	2.1	1.9	:	10.7	10.4	:	2.4	2.6	:
RO	33.9	32.6	:	22.5	19.9	:	7.6	8.4	:	3.8	4.3	:
TR	23.1	:	:	3.7	:	:	2.5	:	:	16.9	:	:
EEA	1 794.5 s	1 830.7 s	1 859.4 s	986.8 s	1 015.2 s	1 030.1 s	258.4 s	252.1 s	254.6 s	528.2 s	543.2 s	553.9 s
IS	2.6 e	2.9	:	1.1 e	1.3	:	0.7 e	0.7	:	0.7 e	0.7	:
NO	:	26.6	:	:	14.4	:	:	4.8	:	:	7.5	:

NB: ACC excludes MT.

(1) **Exceptions to the reference year 2000**
EL: 1999;
FR and AT: 1998;
UK — all sectors: 1993.

Source: Eurostat.

R&D personnel in the EU rose by 1.6% in 2002 compared to 2001. By institutional sector, the annual growth rates in the EU were positive: 2.0% in the Higher education sector – HES, 1.5% in the Business enterprise sector – BES – and 1.0% in the Government sector – GOV. Except for the GOV, the growth rates registered in 2001 were below those of 2001. At the national level, identical trends may be observed for all sectors as all annual growth rates registered were positive but were, for the majority of the countries, below that of the previous year. The highest annual growth rates were reached by Greece (14.4%), ahead of Iceland (10.3%), Belgium and Denmark (5.8% each). Whatever the sector, the most important increases were observed in countries with relative low levels of R&D personnel in volume, with the exception of the United Kingdom in the GOV. In the BES, the Netherlands (26.5%), Greece (17.9%), Iceland (17.3%), Denmark and Portugal (9.0% each), recorded the highest annual growth rates. Whilst in the HES Luxembourg (50.0%) and Greece (18.6%) retained the highest rates, Ireland (14.6%) did the same in the GOV.

Decreases dominated in the Candidate Countries, where the number of R&D personnel diminished between 2000 and 2001 for 6 countries. For all sectors, the highest rates were recorded for Turkey and the Czech Republic with 15.0% and 7.9%, respectively. In the BES, the highest R&D personnel increase was observed in Estonia, where an annual growth rate of 50.1% was registered after an important reduction of R&D personnel in 2000. Turkey (14.0%) and Lithuania (13.5%) also retained high rates in the BES. In Latvia negative growth of 0.9% in 2001 was preceded by a strong increase of 195.7% in 2000. The Government sector is the institutional sector with most negative rates recorded, with 8 countries having seen their R&D personnel reduced. In the HES, only Hungary, Poland and the Slovak Republic saw their R&D personnel reduced, with Turkey recording the highest annual growth rate in this sector (15.6%) – Table 3.3.

Table 3.3.

Annual growth rate of R&D personnel in FTE in %, by institutional sector EU-15, Candidate Countries, Iceland and Norway 2000 to 2002 (1)

	All sectors			Business enterprise			Government			Higher education		
	2000	2001	2002	2000	2001	2002	2000	2001	2002	2000	2001	2002
EU-15	3.1 s	2.0 s	1.6 s	3.6 s	2.9 s	1.5 s	-0.2 s	-2.5 s	1.0 s	3.7 s	2.9 s	2.0 s
BE	6.8 ep	5.8 ep	:	6.8 ep	7.7 ep	1.1 f	56.7 ep	5.3 ep	:	-0.1 ep	2.1 ep	:
DK	5.7 e	5.8	:	8.7 e	9.0	:	-9.3 r	-3.0	:	0.0 r	3.2	:
DE	0.9 e	0.5 e	:	1.9 e	0.6 e	:	-1.1	-0.5 e	:	-0.7	1.1 e	:
EL	14.4	:	:	17.9	:	:	-0.6	:	:	18.6	:	:
ES	18.0 e	4.3	:	22.8 e	-1.3	:	0.5 r	4.8	:	21.8 r	10.4	:
FR	3.1 r	2.2	:	5.5 r	2.5	:	2.3 r	-0.8	:	2.3 r	3.3	:
IE	3.8	:	:	4.8	4.6	:	62.4	14.6 p	:	-15.7	:	:
IT	5.3	:	:	7.3	:	:	1.3	:	:	5.4	:	:
LU	:	:	:	:	:	:	:	13.1	:	:	50.0	:
NL	1.7	:	:	5.2	26.5 p	:	-14.1	-1.4 p	:	10.0	:	:
AT	5.1	:	:	6.2	:	:	0.0	:	:	4.0	:	:
PT	2.9	5.1 e	:	10.5	9.0 e	:	2.4	0.6	:	1.7	5.2	:
FI	4.0	1.6	:	5.6	2.4	:	-2.3 i	-0.3 i	:	4.2	0.9	:
SE	-2.5	4.0	:	-5.5	5.8	:	-5.6	-6.1	:	5.4	1.7	:
UK	:	:	:	-4.8 r	4.3	-3.8 f	0.0 r	-20.9 r	11.5 f	:	:	:
ACC	-0.80			-3.25			-1.25			0.98		:
CZ	0.4	7.9	:	-6.2	4.5	:	2.7	8.7	:	12.6	13.4	:
EE	-18.4	-0.4	:	-32.5	50.1	:	-5.7	-26.3	:	-20.7	0.6	:
CY	-0.1	1.4	:	7.7	-1.0	:	-6.8	1.6	:	9.6	4.8	:
LV	26.7	0.5	:	195.7	-0.9	:	-12.0	-8.5	:	16.4	4.8	:
LT	-7.8	1.3	:	42.6	13.5	:	-10.0	-5.7	:	-8.5	5.9	:
HU	10.3	-2.5	:	9.7	4.8	:	2.8	-5.3	:	18.9	-5.2	:
MT	:	:	:	:	:	:	:	:	:	:	:	:
PL	-4.2	:	:	-8.5	:	:	-1.4	:	:	-3.4	:	:
SI	0.9	0.5	:	-1.8	3.4	:	-2.1	-6.9	:	8.7	2.1	:
SK	2.5	-5.2	:	-9.1	-8.0	:	2.3	-4.8	:	15.7	-3.1	:
BG	-5.1	-2.0	:	-12.5	-11.9	:	-0.6	-2.2	:	-15.4	6.2	:
RO	-23.1	-3.7	:	-29.6	-11.6	:	-13.6	11.2	:	14.0	13.4	:
TR	15.0	:	:	14.0	:	:	12.8	:	:	15.6	:	:
EEA	3.1 s	2.0 s	1.6 s	3.6 s	2.9 s	1.5 s	-0.2 s	-2.4 s	1.0 s	3.7 s	2.8 s	2.0 s
IS	10.7 e	10.3	:	19.4 e	17.3	:	6.3 e	6.0	:	1.6 e	3.5	:
NO	1.0	2.3	:	1.4	3.8	:	-1.0	-0.2	:	1.8	1.2	:

NB: Annual growth rate with regard to the previous year.
ACC excludes MT.

(1) **Exceptions to the reference period 1999-2000**
EL, PT and NO — all sectors, GOV and HES: 1997-99;
FR — all sectors, BES, GOV and HES: 1998-2000;
AT — all sectors, GOV and HES: 1993-98;
SE — all sectors, GOV and HES): 1998-99.
Exceptions to the reference period 2000-2001
PT: 1999-2001.

Source: Eurostat.

R&D personnel in head count — HC

R&D personnel in head count grew at an annual average growth rate of 1.9% in the EU during the 1999-2001 period

According to Eurostat estimations, in 2002 2.5 million people measured in head count worked in R&D in the EU, representing an increase of 1.9% compared to 2001 — Figure 3.6.

The ranking in terms of R&D personnel measured in head count — HC — matches that in FTE. The four leading countries in 2001 included Germany (636 857), France (381 098), the United Kingdom (357 143) and Italy (215 155). The R&D personnel increase during the 1999-2001 period for these countries was quite moderate. Annual average growth rates were positive but below the EU average in Germany (0.6%) and France (1.5%), whereas it was negative for Italy (-3.2%). Among the countries with a high volume of R&D personnel, Spain with 209 011 people, recorded an increase of 8.3% over the previous year. Only Greece, where the highest annual average growth rate was observed (32.2%), and Austria (10.8%), displayed higher figures than Spain among the EU countries. With the exception of Italy, R&D personnel in head count increased in all countries — Figure 3.7.

Figure 3.6. **R&D personnel in thousands of HC all sectors EU-15, Iceland and Norway 2001 (1)**

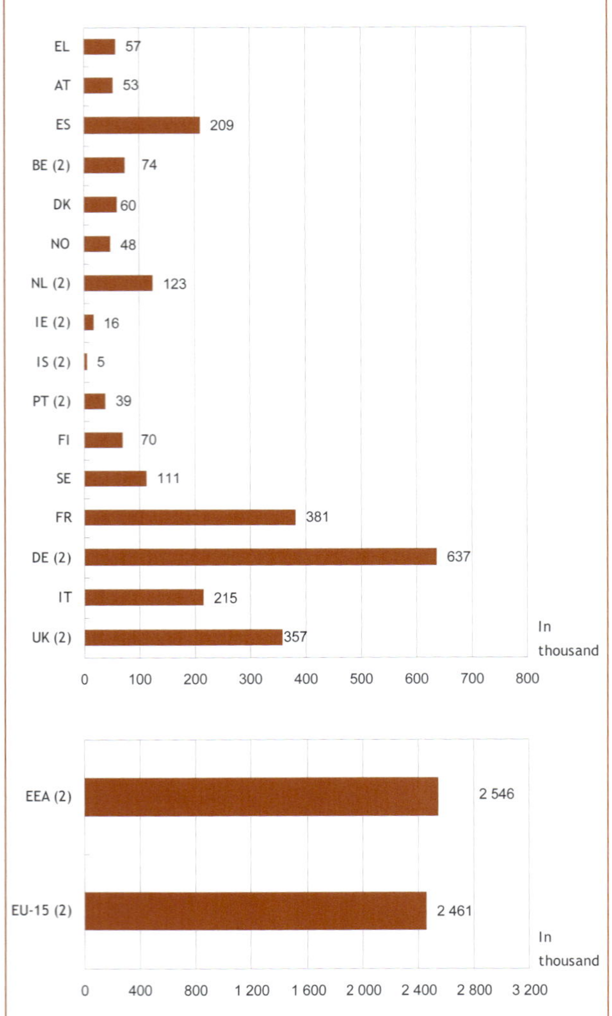

(1) **Exceptions to the reference year 2001**
EU-15, EEA and IS: 2002;
IE and NL: 2000;
EL and IT: 1999;
FR and AT: 1998; UK: 1993.
(2) EU-15, BE, DE, IE, NL, PT, UK and EEA: Eurostat estimates;
IS: estimated value.

Source: Eurostat.

Figure 3.7. **Annual average growth rates of R&D personnel in HC in %, all sectors EU-15, Iceland and Norway 1999-2001 (1)**

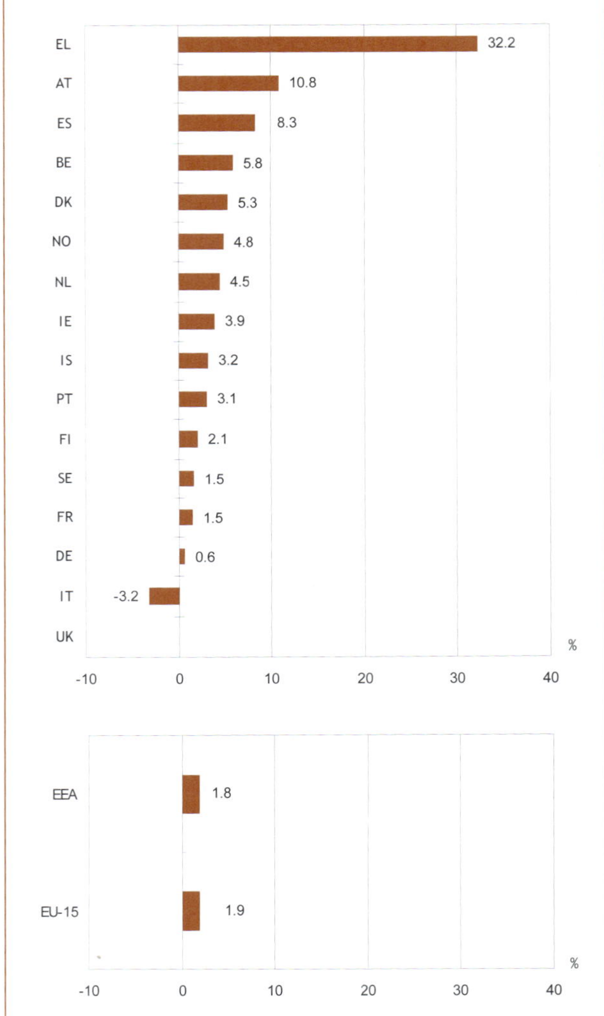

NB: UK can not be calculated as only one year is available.
(1) **Exceptions to the reference period 1999-2001**
EU-15, EEA and IS: 2000-2002; IE and NL:1998-2000;
EL: 1997-99;
IT: 1997-98;
FR: 1996-98;
AT: 1993-98.

Source: Eurostat.

Researchers in full-time equivalent — FTE

The number of researchers increased in all Member States during the 1999-2001 period

In 2002, just over 1 million researchers measured in full-time equivalent — FTE — were employed in the EU, their number having increased by 77 750 since 1999 — Figure 3.8. This positive trend applies also to the national level, as all countries in the EEA, except for Iceland, saw their number of researchers increased between 1999 and 2001.

In terms of volume, most researchers work in Germany — 259 597, France — 177 374 — and the United Kingdom — 158 586. Following these countries is Spain. As opposed to the results for total R&D personnel where Italy's figure is superior to that of Spain, the number of researchers in FTE is, proportionally and in volume, higher in Spain. In terms of growth measured by annual average growth rates, the highest increases were observed in Greece and Spain with rates of 16% and 14%, respectively. Following Greece and Spain are Iceland and the United Kingdom with annual average growth rates above 8%. Germany in turn recorded a very small increase of below 1%.

Figure 3.8. **Researchers in FTE, all sectors EU-15, Iceland and Norway 1999 and 2001 (1)**

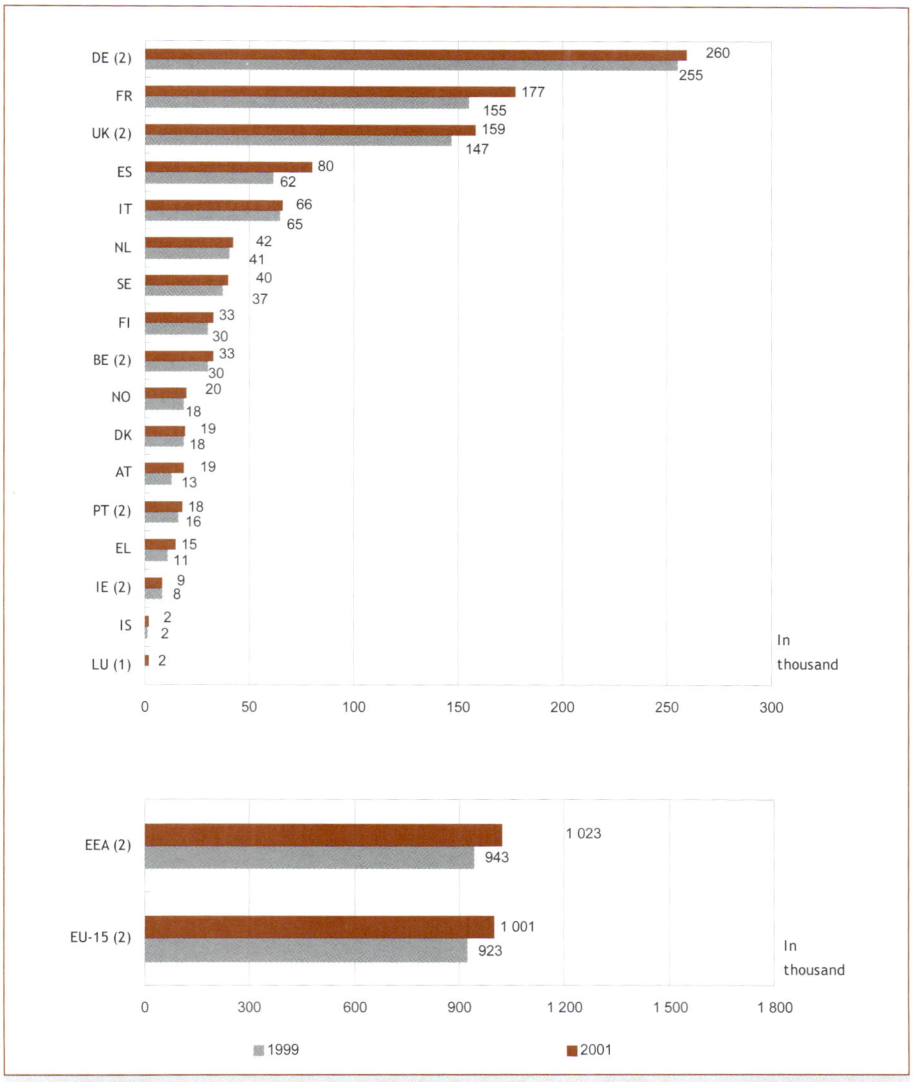

(1) **Exceptions to the reference years 1999 and 2001**
EU-15 and EEA: 1999 and 2002; IE, IT and NL: 1999 and 2000; LU: 2000 (only data for one year is available);
FR: 1998 and 2001; EL and SE: 1997 and 1999,
FI: 1998 and 1999, UK: 1997 and 1998; AT: 1993 and 1998.
(2) BE, DE, IE (1999) and PT (2001): estimated values;
EU-15, UK and EEA: Eurostat estimates.

Source: Eurostat.

Researchers by gender

Women researchers still under-represented in the EU

Women researchers are still under-represented in the EU compared to men, especially in the Business enterprise sector – BES. In 2001, they accounted for 44% of total researchers in Portugal and 39% in Iceland where the maximum values were obtained among EEA countries. For all sectors, the lowest proportion of female researchers in the EEA were observed in Germany (24%), Belgium and Norway (28% each). For all Member States, female researchers are more rare in the Business enterprise sector – BES. In the EEA, the highest share of women amongst researchers in the BES was retained by Iceland with 37 %, whilst there were less than one out of three women researchers in the rest of the countries.

The percentage of women among researchers is in general higher among the Candidate Countries. Indeed, parity is reached in Latvia in both all sectors (55%) and BES (59%) and rates above 45% are registered in all sectors for Lithuania and Bulgaria. The lowest proportion of female researchers for all sectors in the Candidate Countries was observed in the Czech Republic (26%). Compared to the EU, the differences between the percentage of female researchers in all sectors and in the BES are not as large in the Candidate Countries. In fact, for five Candidate Countries – Romania, Lithuania, Cyprus, Latvia and Bulgaria – the rates for all sectors and the BES were very close – Figure 3.9.

Figure 3.9. **Percentage of female researchers in FTE (1), all sectors and BES EU-15, Candidate Countries, Iceland and Norway 2001 (2)**

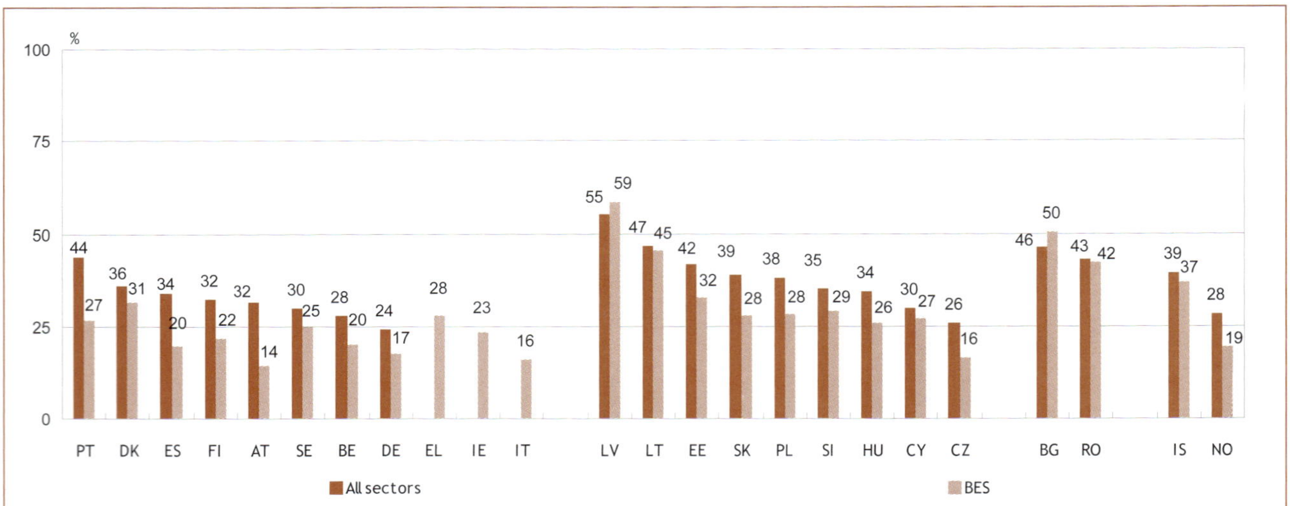

(1) **Exceptions to data in full-time equivalent — FTE**
LU, FI, HU, PL and IS: data in head count — HC.

(2) **Exceptions to the reference year 2001**
UK: 2002;
DE, CZ, EE, CY, HU, PL, SI, SK, BG and RO: 2000;
EL: 1999;
IT and AT: 1998.

Source: Eurostat.

Researchers by field of science

Medical and social sciences are the fields with the highest proportion of researchers employed

For most countries and fields of science, the proportion of researchers was superior in the Higher education sector — HES — than in the Government sector — GOV. Within the same country, important disparities could be observed across fields of science. The situation was the same for both the GOV and the HES by field of science, where the top researcher proportions were obtained in medical sciences and social sciences. At the national level, Germany displayed the lowest ratios, whereas Portugal retained the highest values. In the GOV for instance, 87% of Portugal's R&D personnel in medical sciences were researchers against only 45% in Germany. The situation was similar in the social sciences where the gap observed was also close to 40%. Similar trends were observed in the HES where the proportion of researchers also varied highly between the maximum and minimum values. For both medical sciences and social sciences, the gap between the highest and the lowest countries reached 50% and 42%, respectively.

Compared to total R&D personnel, very high proportions of researchers are observed for Candidate Countries in social sciences and medical sciences in the GOV on the one hand and in social sciences and humanities in the HES on the other. In general, the percentage of researchers is higher in the HES than in the GOV. Apart from a few exceptions, the share of researchers was above 50% both in the GOV and in the HES — Table 3.4.

Table 3.4. **Researchers in FTE as a % of total R&D personnel by field of science, GOV and HES**
EU-15, Candidate Countries, Iceland and Norway
2001 (1)

	DK	DE	ES	IE	NL	PT	SE	CZ	EE	CY	LV	LT	HU	PL	SI	SK	BG	RO	IS	NO
Government																				
Natural sciences	73	53	56 e	:	:	58 e	:	60	61	34	56	77	61	58	59	61	68	71	:	62
Engineering and technology	72	58	46 e	:	:	60 e	:	62	61	80	39	65	70	56	35	67	59	66	:	52
Medical sciences	58	45	73 e	:	:	87 e	:	61	57	73	74	59	46	65	89	54	66	54	:	71
Agricultural sciences	58	40	48 e	:	:	42 e	:	66	48	15	63	45	39	50	45	45	39	54	:	52
Social sciences	76	58 i	57 e	:	:	97 e	:	61	90	58	63	85	72	67	67	74	60	74	:	83
Humanities	79	:	64 e	:	:	79 e	:	67	60	19	10	76	59	80	84	47	79	88	:	73
Total	**64**	**53**	**57**	**:**	**:**	**60 e**	**:**	**62**	**59**	**23**	**52 r**	**52**	**57**	**59**	**58**	**60**	**60**	**69**	**:**	**65**
Higher education																				
Natural sciences	68	74	85 e	83	59	99	80	79	74	87	67	90	63	77	89	78	79	72	78	76
Engineering and technology	83	70	86 e	69	63	83	74	75	86	91	70	93	70	80	75	86	67	61	78	78
Medical sciences	53	36	86 e	70	67	66	65	48	71	:	82	81	54	82	54	80	93	87	68	70
Agricultural sciences	60	62	86 e	45	64	71	70	42	69	:	55	70	45	84	88	:	60	60	71	64
Social sciences	86	85	87 e	96	54 i	91	86	74	82	95	89	96	81	87	76	96	83	94	68	81
Humanities	91	87	88 e	97	:	95	89	71	88	100	89	98	86	91	87	98	92	46	73	83
Total	**74**	**66**	**86**	**83**	**59**	**88**	**76**	**71**	**78**	**93**	**74 r**	**79**	**66**	**83**	**77**	**85**	**78**	**67**	**73**	**76**

(1) **Exceptions to the reference year 2001**
BE, DE and SE: 1999;
IE, NL, PT (GOV), CZ, EE, CY, HU, PL, SI, SK, BG and RO: 2000.

i DE (GOV) and NL (HES): social sciences includes humanities.

Source: Eurostat.

3.4. R&D personnel in the European regions

Leading regions in R&D personnel

In 2001, Île de France led in absolute terms, whereas Uusimaa (Suuralue) retained the highest rate as a % of labour force

In 2001 one quarter of the EEA's R&D personnel in full-time equivalent – FTE – was concentrated in the 10 leading regions. Accounting for 7% of the EEA's total, Île de France (FR) was the leading region in terms of R&D personnel in FTE. German regions were the most represented among the top 10, with Oberbayern and Stuttgart laying second and third, respectively. Besides Germany and France, only three other countries had regions in the top 10 ranking: Denmark, which is classified as a region at NUTS level 2, Italy and Spain – Figure 3.10.

Figure 3.10. **Percentage of R&D personnel accounted for by the top 10 EEA regions in FTE, all sectors 2001**

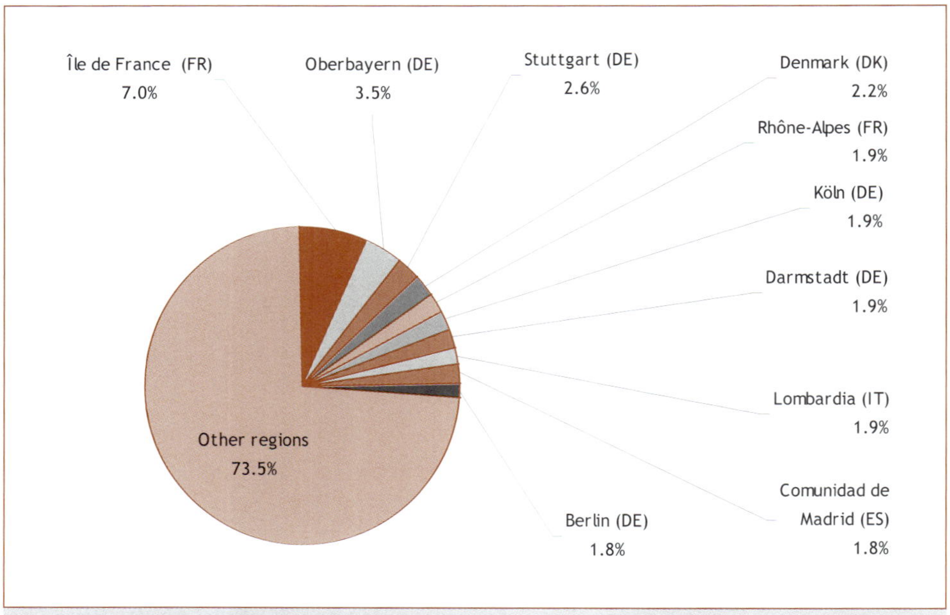

NB: Data for all sectors are available for the following countries: DE, EL, ES, FR, IT, NL, AT, PT, FI, SE and IS. According to the NUTS classification, for Denmark the entire national territory is considered as a NUTS 0, 1 or 2 region.

(1) **Exceptions to the reference year 2001**
FR and IT: 2000;
DE: 1997.

Source: Eurostat.

In the Candidate Countries, a slightly different picture can be observed for countries for which data are available even if one country dominates the top 10. Indeed, four Polish regions are included in the 10 first regions for R&D personnel. Mazowieckie (PL) stands out at the first position followed by Kozep-Magyarorszag (HU), Lithuania, which is classified at the NUTS 2 level, Yugozapaden (BG) and Praha (CZ) — Figure 3.11.

As regards R&D personnel as a % of labour force in the EEA, the Finnish region of Uusimaa (Suuralue) was leading in 2001 with 3.86% of its workforce employed in R&D. Following Uusimaa were Stockholm (SE) with 3.72% and two German regions Oberbayern (3.44%) and Braunschweig (3.33%). These three countries alone accounted for 8 regions among the top 10. The share of R&D personnel in the labour force reached by the top 10 regions was at least 1.2 percentage points above the EU average. However, a gap of more than 1 percentage point is observed between the first region Uusimaa (Suuralue, FI) and the 10th region Östra Mellansverige (SE) — Table 3.5.

Figure 3.11. **Top 10 regions in R&D personnel in FTE, all sectors Candidate Countries 2001**

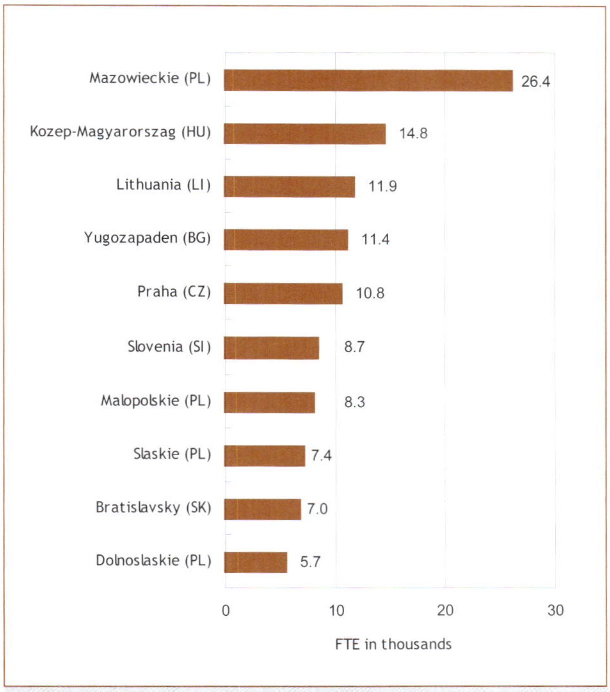

NB: Data for all sectors are available for all Candidate Countries **except**: MT, RO and TR.
For EE, CY, LV, LT and SI the entire national territory is considered as a NUTS 0, 1 and 2 region, which explains their appearance in the regional rankings.

Source: Eurostat.

Table 3.5. **Top 10 regions in R&D personnel as a % of labour force, all sectors EEA 2001**

Region	Country	R&D personnel as a % of the labour force
EU-15		1.39
Uusimaa (Suuralue)	FI	3.86
Stockholm — 1999	SE	3.72
Oberbayern — 1997	DE	3.44
Braunschweig — 1997	DE	3.33
Pohjois-Suomi	FI	3.24
Wien — 1998	AT	3.14
Iceland — 2002	IS	3.09
Övre Norrland — 1999	SE	2.87
Stuttgart — 1997	DE	2.73
Östra Mellansverige — 1999	SE	2.73

Source: Eurostat.

Regional disparities in R&D personnel

As a percentage of labour force Uusimaa (Suuralue) is leading

Those gaps are also noticeable between countries when R&D personnel is measured as a % of labour force. Almost 2.3 percentage points separate the top region of Finland, Uusimaa (Suuralue), from the top region of Portugal, Lisboa e Vale do Tejo. Besides Finland, the top region for three other countries – Sweden, Germany and Austria – is above 3%. At the opposite end, even Portugal's top region is under the EU average (1.39%). Regions with the lowest proportion of R&D personnel by countries show quite similar figures from one country to another. As a consequence, the gaps between the top and the lowest region vary in large proportions among the Member States – Figure 3.12.

A similar picture as for total sectors, is shown in the BES for R&D personnel as a % of labour force. For instance, with 7 regions, among which Oberbayern and Stuttgart lead, Germany has a majority of the top 15 regions. This leadership is shared with Finland (3 regions) and Sweden (2 regions) – Map 3.1.

Figure 3.12. **Regional disparities in R&D personnel as a % of labour force, all sectors EU-15, Iceland and Norway 2001 (1)**

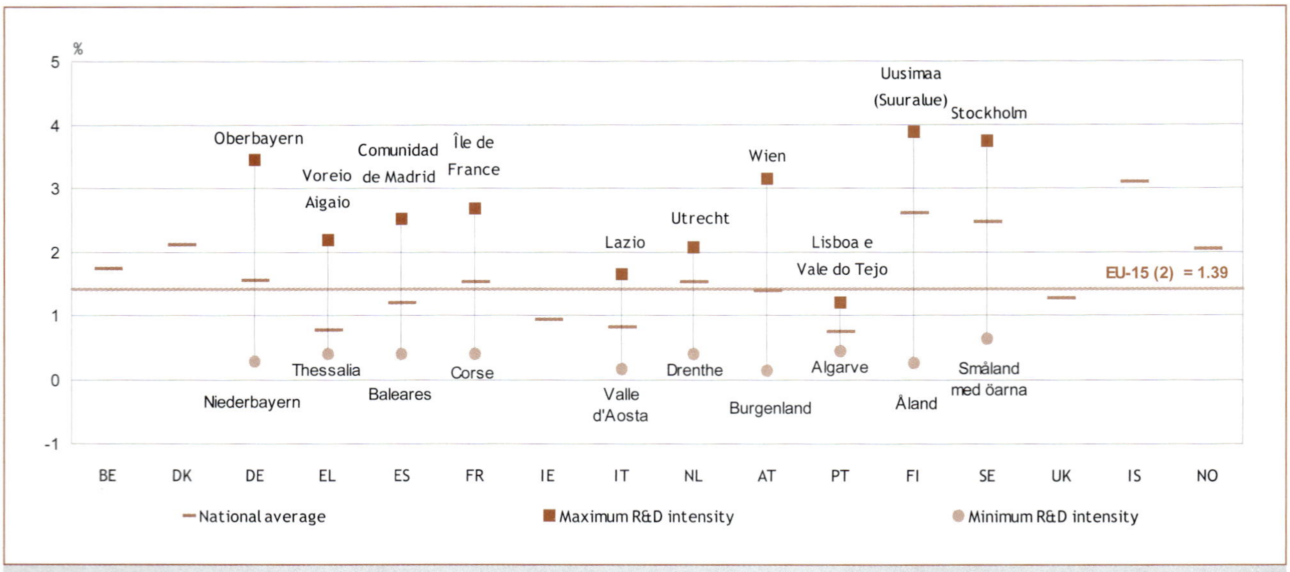

(1) **Exceptions to the reference year 2001**
EU-15: 2002; FR, IE and NL: 2000;
SE: 1999;
AT: 1998;
DE and EL: 1997;
UK: 1993.
(2) EU-15: Eurostat estimate.

Source: Eurostat.

Map 3.1.

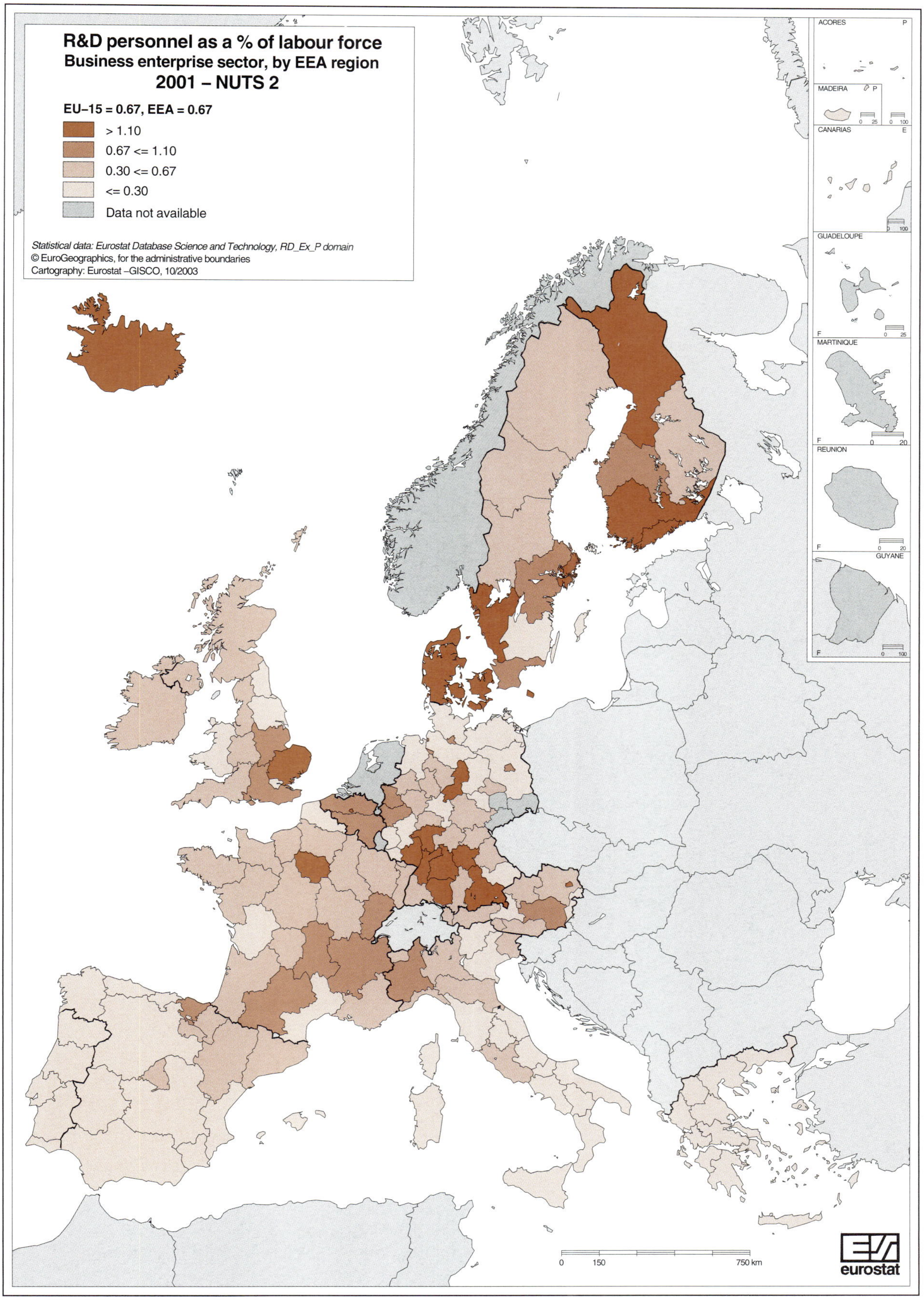

EU-15 = 0.67% and EEA = 0.67%
refer to the EU-15 and the EEA averages, i.e. in 2001 and in the Business enterprise sector — BES, R&D personnel in the EU and the EEA accounted to 0.67% of their respective labour force.

Exceptions to the reference year 2001
IT: 2000; SE: 1999;
FR and AT: 1998; DE and EL: 1997.

Exceptions to data at the NUTS 2 level
BE, IE and UK: classified at NUTS 1 level.

4.1. Introduction

Co-ordinated sets of data on Human Resources devoted to Science and Technology can, when linked to demographic statistics, be used to review the supply of, and demand for, science and technology personnel. This, with a view to evaluating the consequences for future research and industrial performance, planning education and training, and measuring the diffusion of knowledge incorporated in human resources.

In essence, human resources in science and technology, or HRST for short, are people that are highly qualified. HRST have a number of different sub-categories, which are most easily understood by looking at Figure 4.1. They are measured following international standards, using the individuals' level of formal education, their occupation type, or both ([1]).

Figure 4.1. shows that HRST fulfil at least one of the following conditions:

- successfully completed education at the third level in a S&T field of study ([2]) — ISCED '97 version levels 5a, 5b or 6, *or*
- are not formally qualified as above but are employed in a S&T occupation where the above qualifications are normally required — ISCO '88 COM codes 2 or 3.

Even though the official definition of HRST as shown in the *Canberra Manual* ([3]) contains the terms 'S&T', these terms do not restrict the definition: HRSTE covers all fields of study i.e. anybody who successfully completed third level education; HRSTO refers to two specific major ISCO classes that are broader than what one might expect from scientific and technological activities in a stricter sense (notably ISCO 2 'Professionals' and ISCO 3 'Technicians and associate professionals') — see methodological notes.

An HRST sub-set of particular interest is scientists and engineers. Those more likely to be involved in leading-edge technology professions are 'Physical, mathematical and engineering' occupations (ISCO '88 COM code 21), as well as 'Life science and health' occupations (ISCO '88 COM code 22) ([4]).

The data and indicators presented in this chapter are, for the most part, prepared in line with the recommendations laid down in the Manual on *The Measurement of Human Resources devoted to S&T — Canberra Manual*.

The data are taken from two principal sources:

- Indicators concerning the stocks of human resources in science and technology use data from the *European Union Labour Force Survey* — EU LFS.
- The inflows, meanwhile, use data from Eurostat's education database, collected via the joint Unesco/OECD/Eurostat — UOE — questionnaire on education statistics.

The education inflows in Chapter 4.2. are a useful measurement of the current and future supply of HRST. They are named as such since upon achieving an education at the tertiary level, an individual becomes HRST i.e. moves into the stock of HRST. Inflows can be sub-divided into various groups, each providing a different level of focus. Measurements themselves can be divided into participation in and graduation from tertiary education, with the former used to estimate potential future inflow rates into the labour market and the latter the actual inflows. Additional focus is provided on the most highly educated individuals, that is recipients of PhD level awards.

The data on stocks in Chapter 4.3., meanwhile, provide an indication of how many HRST there are at a particular point in time. These can then be broken down to provide information on socio-economic categories of interest, such as the ratio of men to women, the importance of age or the sector of activity in which people are more likely to work.

(1) Education data follow the ***International Standard Classification for Education*** — ISCED, whilst occupation data follow the ***International Standard Classification for Occupation*** — ISCO.

(2) Note that according to the ***Canberra Manual***, the seven broad S&T fields of study are 'Natural Sciences', 'Engineering and Technology', 'Medical Sciences', 'Agricultural sciences', 'Social sciences', 'Humanities', 'Other fields', ***Canberra Manual***, § 71.

(3) ***Manual on the Measurement of Human Resources devoted to S&T — Canberra Manual***, OECD, Paris, 1994.

(4) Readers should note that scientists and engineers differ from the ***Frascati Manual*** definition of researchers, which includes persons in ISCO-88 Major Group 2 'Professional Occupations' plus 'Research and Development Department Managers — ISCO-88 1237;
Standard method for surveys on R&D and experimental development — Frascati Manual, OECD 2002, paragraph 312.

Figure 4.1. **Categories of HRST**

<table>
<tr><td></td><td></td><td></td><td colspan="3">HRSTE
– Education –</td><td></td></tr>
<tr><td></td><td></td><td></td><td colspan="3">Tertiary education</td><td>Lower than tertiary education</td></tr>
<tr><td></td><td></td><td></td><td>ISCED 6</td><td>ISCED 5a</td><td>ISCED 5b</td><td>ISCED < 5</td></tr>
<tr><td rowspan="2">HRSTO
– Occupation –</td><td>ISCO 2</td><td>Professionals</td><td colspan="3" rowspan="2">HRST core – HRSTC</td><td rowspan="2">HRST without tertiary education</td></tr>
<tr><td>ISCO 3</td><td>Technicians</td></tr>
<tr><td></td><td>ISCO 1</td><td>Managers</td><td colspan="3" rowspan="2">HRST non-core</td><td rowspan="2"></td></tr>
<tr><td></td><td>ISCO 0, 4-9</td><td>All other occupations</td></tr>
<tr><td></td><td></td><td>Unemployed</td><td colspan="3">HRST unemployed – HRSTU</td><td>Non-HRST unemployed – NHRSTU</td></tr>
<tr><td></td><td></td><td>Inactive</td><td colspan="3">HRST inactive</td><td></td></tr>
</table>

Source: Eurostat.

4.2. Education inflows

Participation in tertiary education

Over ten million people in the EU were following tertiary education courses in 2001, equivalent to a little over a quarter of all people aged 20-29 – Table 4.1. With a similar proportion of 20-29 year olds, the Acceding Countries had just over 2.9 million students in tertiary education in 2001.

Moreover, these student numbers are generally growing – Figure 4.2. Between 1998 and 2001 the number of people following a tertiary level education grew at an annual average rate of 1.2% in the EU and 12.4% in the Acceding Countries. Sweden saw the highest EU growth over this period at 8.4%, in contrast to the contraction in student numbers in Luxembourg, Austria, Italy and to a lesser extent Germany (Luxembourg does not have a complete university system – most students study abroad). Meanwhile, all Acceding Countries saw higher enrolment levels. Indeed, all of the Acceding Countries but Cyprus had higher or equal growth rates compared to the highest evident in the EU (Sweden).

As shown in Table 4.1., science and engineering courses together accounted for just over a quarter of all tertiary studies in the EU in 2001, though engineering was marginally more popular at 14.5% of all courses than science (11.8 %). This trend was reflected in most EU Member States for which data were available, the exceptions being Ireland, Luxembourg and the UK – though it is worth underlining that engineering in Ireland was still as popular as in the EU as a whole. Furthermore, it was in Ireland that the highest proportion of students studying science could be found (20.6 %), followed by the UK (17.4 %).

Table 4.1. **Participation in tertiary education, in total and selected fields of study by sex in comparison to the population aged 20-29 EU-15, Candidate Countries, Iceland and Norway 2001 (1)**

	Participation in tertiary education in 2001						
	Total participation			In science		In engineering, manufacturing and construction	
	Total	% population aged 20-29	% women in total	Total	% of total participation	Total	% of total participation
EU-15 (2)	**10 335 634**	**26.0**	**53.2**	**1 223 512**	**11.8**	**1 498 225**	**14.5**
BE	359 265	27.3	52.8	35 157	9.8	40 886	11.4
DK	190 791	27.2	56.5	19 359	10.1	20 277	10.6
DE	2 083 945	23.7	48.7	282 960	13.6	323 953	15.6
EL	:	:	:	:	:	:	:
ES	1 833 527	28.1	52.5	237 402	13.0	303 122	16.6
FR	:	:	:	:	:	:	:
IE	166 600	25.6	54.7	26 683	20.6	19 343	14.9
IT	1 812 325	23.1	56.0	135 668	7.5	299 778	16.6
LU	2 533	4.5	:	245.0	9.7	181	7.1
NL	504 042	24.6	50.5	28 818	5.7	53 641	10.7
AT	289 722	28.6	:	33 480	11.6	40 448	14.0
PT	387 703	23.6	57.0	27 671	7.1	79 006	20.4
FI	279 628	44.7	53.9	30 472	10.9	72 303	25.9
SE	358 020	32.6	59.1	38 971	10.9	68 206	19.1
UK	2 067 349	27.8	54.5	360 106	17.4	217 529	10.5
ACC (2)	**2 916 821**	**25.5**	**56.8**	**177 463**	**6.1**	**420 969**	**14.4**
CZ	260 044	15.5	50.1	36 338	14.6	41 536	16.7
EE	57 778	30.3	60.1	5 011	8.7	7 320	12.7
CY	11 934	14.4	58.0	1 562	13.1	550	4.6
LV	102 783	31.7	61.8	6 592	6.4	10 128	9.9
LT	135 923	25.5	59.8	6 716	4.9	29 419	21.6
HU	330 549	20.6	54.8	16 011	4.8	51 256	15.5
MT	7 422	:	54.8	358	4.8	459	6.2
PL	1 774 985	30.6	58.0	89 143	5.5	234 638	14.4
SI	91 494	30.5	56.1	4 588	5.0	16 026	17.5
SK	143 909	16.1	51.3	11 144	7.7	29 637	20.6
BG	247 006	22.7	56.3	11 916	4.8	52 777	21.4
RO	533 152	15.9	53.5	26 662	5.3	108 672	21.6
TR	1 091 805	:	40.5	113 673	10.4	211 449	19.4
IS	10 184	25.3	62.7	1 303	12.8	606	6.0
NO	190 054	32.3	59.2	22 841	12.8	12 386	6.9

NB: EU-15 excludes EL and FR.

(1) **Exception to the reference year 2001**
AT: 2000.

(2) EU-15 and ACC: Eurostat estimates.

Source: Eurostat, UOE questionnaire and EU LFS — spring data.

At 14.4%, engineering was just as popular in the Acceding Countries as in the EU and even represented more than 20% of all tertiary education courses in Lithuania and the Slovak Republic. Science however suffered from lower interest than engineering, but to a greater extent in the Acceding Countries than in the EU. On average, 6.1% of all tertiary courses in the Acceding Countries in 2001 were in science. In fact, apart from in Cyprus and the Czech Republic (13.1% and 14.6%, respectively), fewer than one in ten students studied science in the remaining Acceding Countries.

Though women accounted for more than half of all students in practically every country – Turkey and, to a lesser extent, Germany being the exceptions – this was not the case when it came to studying science. Indeed, as is shown in Figure 4.3., in the EU, parity was only achieved in Italy and Portugal; two countries that showed student participation in science well below the EU average. At the EU level, nearly four in every ten students studying science were women in 2001, but this fell to as low as one in four in the Netherlands. In Romania and Bulgaria, however, there were more female than male science students.

Engineering courses have even more problems attracting women. At 37%, Bulgaria had the highest ratio of women engineering students, followed by Lithuania with 31%. In the EU, women accounted for the highest proportions of engineering students in Sweden and Portugal (29% and 28%, respectively).

Figure 4.2.

Annual average growth rates in the number of tertiary education students
EU-15, Candidate Countries, Iceland and Norway
1998 to 2001 (1)

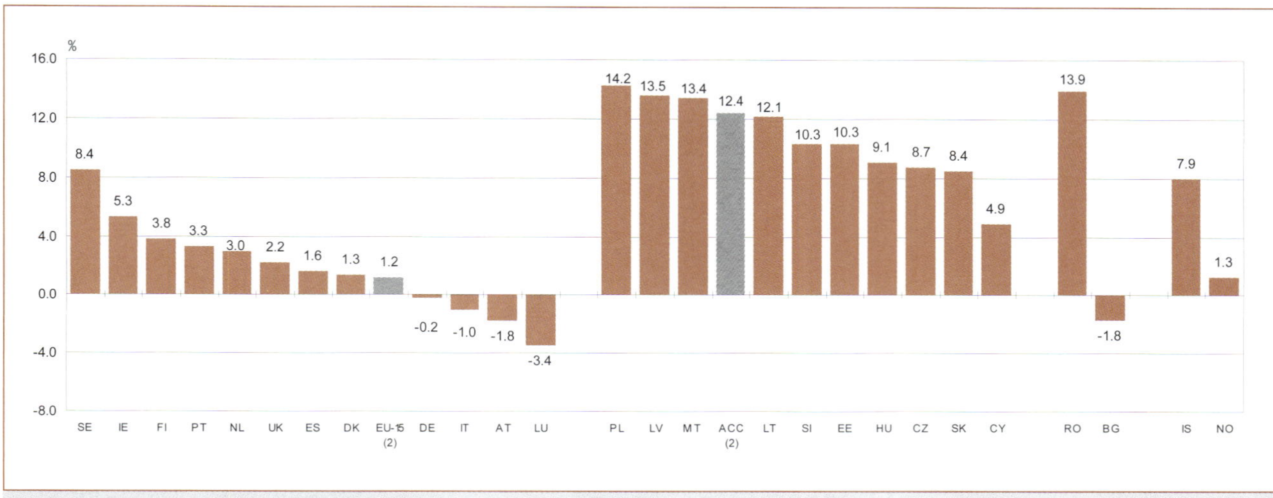

(1) **Exceptions to the reference period 1998-2001**
LU, CY and MT: 1999-2001; AT: 1998-2000.
(2) EU-15 and ACC: Eurostat estimates.

Source: Eurostat, UOE questionnaire.

Figure 4.3.

Proportion of female S&E tertiary students
EU-15, Candidate Countries, Iceland and Norway
2001 (1)

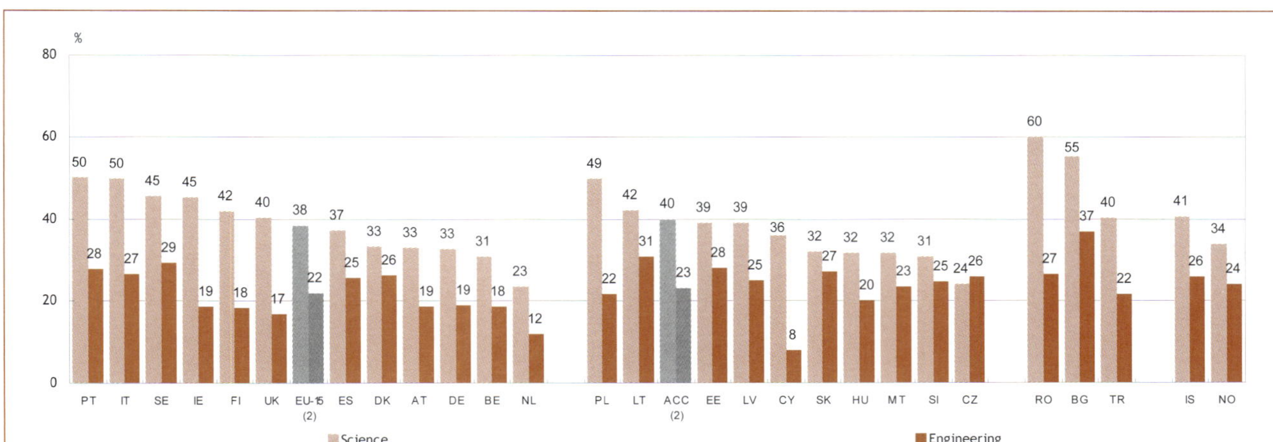

(1) **Exception to the reference year 2001**
AT: 2000.
(2) EU-15 and ACC: Eurostat estimates.

Source: Eurostat, UOE questionnaire.

Figure 4.4. **Interest in S&T developments by areas EU average in % and deviation from EU average per country EU-15**

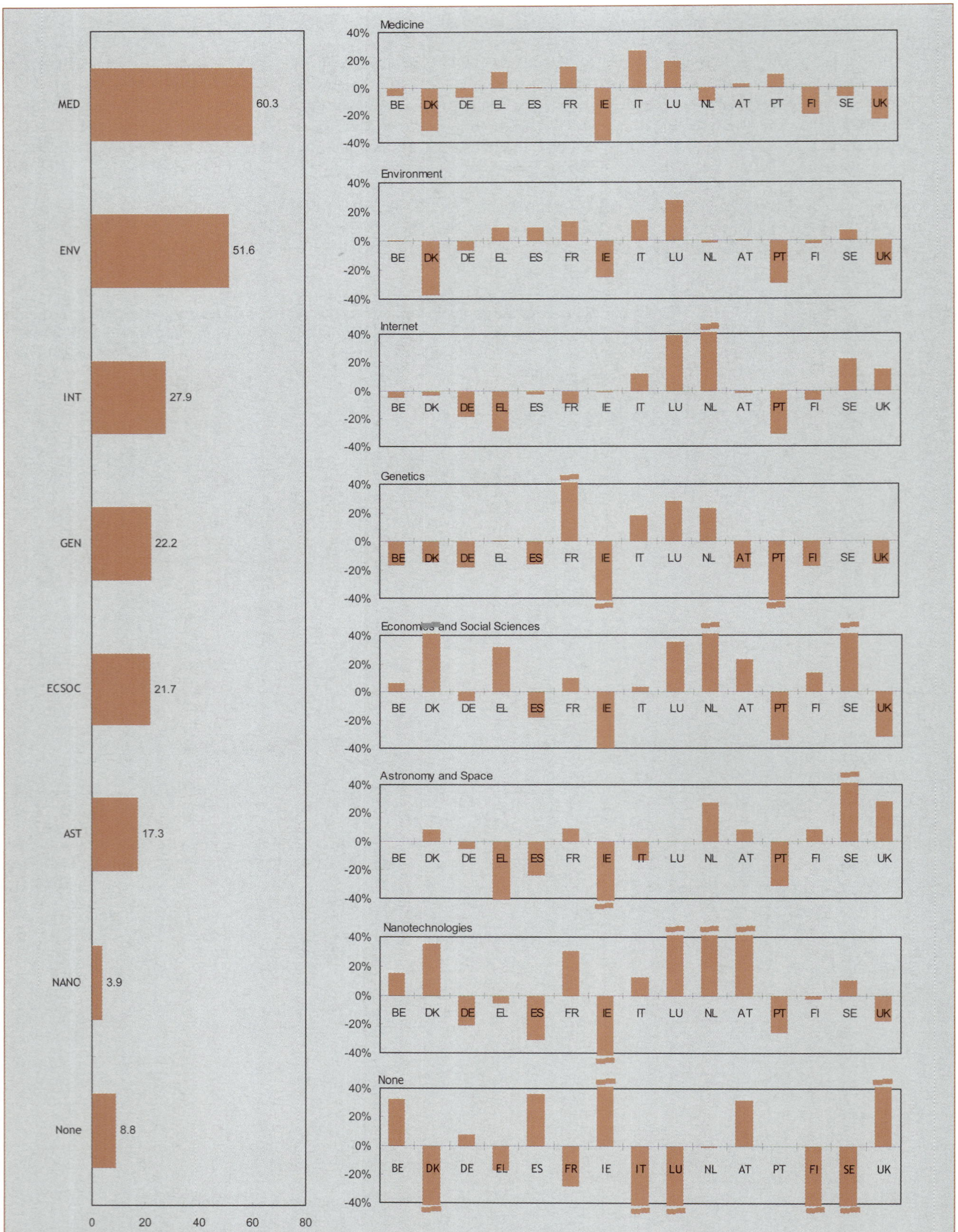

NB: The deviation from the EU average is calculated by: (% country - % EU average) / (% EU average).
Deviations higher than 40% or below - 40 % are not shown in this figure.

Source: *Third European Report on Science and Technology Indicators*, Directorate-General for Research, 2003, p. 199.
Data: Eurobarometer 55.2, Table 3.

General interest in scientific and technological developments

Figure 4.4. details interest in scientific and technological developments among the EU population. As is shown, developments in medicine and the environment are of by far the most interest to the EU population, with 60.3% of EU respondents saying that medicine is of interest to them compared with 51.6% for the environment. The Internet, genetics and economic and social science fall much further behind at between 27.9% and 21.7%.

Large differences exist between the Member States, however. Italians show the most interest in medicine and Luxembourgers in the environment. Again, citizens from Luxembourg and also from the Netherlands show the most interest in the Internet, developments for which the Portuguese show the least interest. The French are the most interested in genetics, as are the Swedish in Astronomy. The data show the Irish and the British to be the least interested in S&T developments.

Figure 4.5. **Participation of foreign students in tertiary education total and share of science and engineering students EU-15, Candidate Countries, Iceland and Norway 2001 (1)**

	Foreign students in any field	Foreign students in S&E fields		Share of foreigners in total S&E students (%)
	Total	Total	%	
BE	38 150	5 691	17.7	7
DK	12 586	2 930	23.3	7
DE	199 132	61 684	32.6	10
EL	:	:	:	
ES	40 689	:	:	
FR	:	:	:	
IE	8 207	:	:	
IT	29 228	5 433	18.6	1
LU	:	:	:	
NL	16 589	3 073	18.6	4
AT	35 891	8 782	24.5	12
PT	14 202	3 460	24.4	3
FI	6 288	2 378	37.8	2
SE	26 304	8 250	31.4	8
UK	225 722	74 047	32.8	13
CZ	7 750	1 997	25.8	3
EE	863	:	:	
CY	2 472	331	13.4	16
LV	7 917	180	2.3	1
LT	628	94	15.0	0.3
HU	11 242	2 290	20.4	
MT	341	25	7.3	3
PL	6 659	580	8.7	0.2
SI	864	235	27.2	1
SK	1 690	217	12.8	1
BG	8 130	1 205	15.0	2
RO	11 669	961	8.4	1
TR	16 656	3 492	21.0	1
IS	421	53	12.6	
NO	8 857	2 045	25.6	

(1) **Exception to the reference year 2001**
AT: 2000.

Source: Eurostat, UOE questionnaire.

International student mobility

Science and engineering more popular among internationally mobile students in countries belonging to EU

National figures for overall participation in tertiary education also include foreign students, defined according to the citizenship of the individual. Though overestimation of non-national students may exist in some countries where permanently resident second generation migrants with foreign nationalities constitute an important group of students, foreign students can otherwise be interpreted as internationally mobile students. Figure 4.5. shows how many foreign students chose to study science and engineering related subjects in 2001, as a proportion of both total foreign students and the number of students studying S&E overall.

In Germany, Austria, the United Kingdom and Cyprus, 10% or more of all students studying S&E in 2001 were foreign. Furthermore, in Germany and the United Kingdom, a third of all internationally mobile students followed science and engineering related disciplines (32.6% and 32.8%, respectively), just below the leading country, Finland (37.8%). This proportion exceeded the popularity of S&E programmes at the national level overall — recall Table 4.1., though by just 3% and 5%, respectively.

In fact, on the whole in the EU there was little difference in the popularity of S&E courses between foreign students and total students, something that was not reflected in the Acceding Countries. Apart from in Cyprus, foreign students represented a low proportion of total S&E students in the ACC. Rather, the internationally mobile students were more likely to be studying subjects other than science and engineering, as indicated by the popularity of S&E amongst foreign students — 25.8% in the Czech Republic — compared to the national average of 31.3% — from Table 4.1.

PhD students

A doctor of philosophy – PhD – is an example of an advanced research degree. They usually require 3-5 years of research and course work, generally after a Master's degree. In that sense, indicators on the number of PhD students provide an idea of the degree to which countries will have researchers of the highest level. Even excluding Germany, Greece, France and Luxembourg, for which no data are available, there were over 250 000 PhD students in the EU in 2001 – Table 4.2. The Acceding Countries, meanwhile, had close to 63 000 people following doctorate level studies.

It is noticeable that the relative popularity of science increased markedly compared to overall tertiary level courses: science was taken by one in five Acceding Country PhDs whereas only by one in 16 Acceding Country tertiary students, in general – recall Table 4.1. In the EU, Ireland had the highest proportion of its doctoral students taking science courses at 46.2%, followed by Belgium and the United Kingdom (34.0% and 31.3%, respectively).

Engineering was less popular, in the EU at least, where as a proportion of all PhD enrolments it ranged from one in four in Finland and one in five in Italy and Sweden to one in 13 in Spain. Around 20% of all doctorate studies in 2001 were in engineering in the Acceding Countries.

Figure 4.6. shows that, in general in the EU Member States, women account for a higher proportion of engineering PhD students than they do when all tertiary level courses are taken into account – recall Figure 4.3. For science, this result is not so clear cut, since women account for more PhD science students in only half of the EU countries for which data are available. Women account for a higher proportion of PhD science students than when taking all tertiary level courses into account in all Acceding Countries but Poland.

Figure 4.6. **Proportion of female S&E ISCED level 6 (PhD) students EU-15, Candidate Countries, Iceland and Norway 2001 (1)**

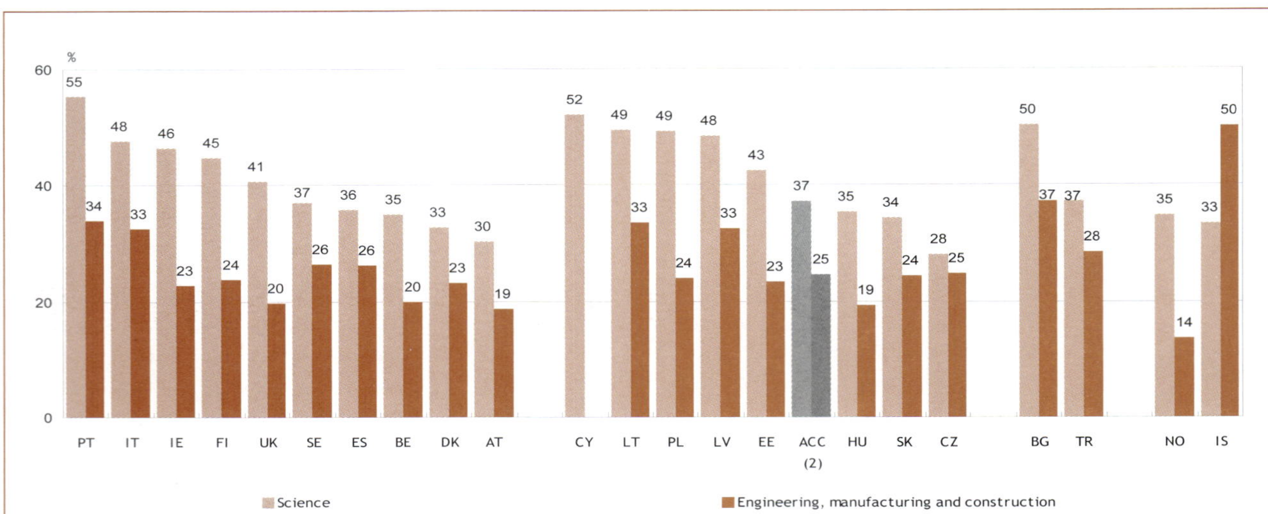

(1) **Exception to the reference year 2001**
AT: 2000.
(2) ACC: Eurostat estimates.

Source: Eurostat, UOE questionnaire.

Table 4.2.

Participation in ISCED level 6 (PhD) education, in total and selected fields of study by sex in comparison to the population aged 20-29 EU-15, Candidate Countries, Iceland and Norway 2001 (1)

	Participation at ISCED 6 level (PhDs) in 2001						
	Total participation at ISCED 6 level (PhDs)			In science		In engineering, manufacturing and construction	
	Total	per 1000 population aged 20-29	% women in total	Total	% of total participation at ISCED level 6	Total	% of total participation at ISCED level 6
EU-15	:	:	:	:	:	:	:
BE	5 613	4.3	36.0	1 911	34.0	731	13.0
DK	3 794	5.4	42.2	795	21.0	557	14.7
DE	:	:	:	:	:	:	:
EL	:	:	:	:	:	:	:
ES	62 530	9.6	39.6	9 299	16.0	4 576	7.8
FR	:	:	:	:	:	:	:
IE	3 059	4.7	45.2	1 349	46.2	351	12.0
IT	20 966	2.7	51.2	3 708	18.5	4 124	20.6
LU	:	:	:	:	:	:	:
NL	7 768	3.8	43.0	:	:	:	:
AT	23 558	23.8	44.0	3 647	15.6	3 560	15.2
PT	12 073	7.3	54.3	2 163	17.9	1 567	13.0
FI	20 631	33.0	48.2	2 998	14.5	5 377	26.1
SE	20 679	18.8	44.3	4 076	19.7	4 561	22.1
UK	75 334	10.1	41.9	23 599	31.3	11 153	14.8
ACC (2)	62 734	5.5	41.5	12 961	20.7	12 843	20.5
CZ	17 719	10.5	35.5	5 298	29.9	3 872	21.9
EE	1 447	7.6	56.1	355	24.5	141	9.7
CY	72	0.9	:	25	34.7	:	:
LV	1 254	2.4	55.5	217	17.3	295	23.5
LT	2 057	6.3	55.4	293	14.2	448	21.8
HU	6 752	4.2	40.7	1 604	23.8	789	11.7
MT	32	:	:	2	6.3	1	3.1
PL	25 622	4.4	44.2	3 855	15.0	5 703	22.3
SI	:	:	:	:	:	:	:
SK	7 779	8.7	38.3	1 312	16.9	1 594	20.5
BG	3 414	3.1	49.6	664	19.4	799	23.4
RO	:	:	:	:	:	:	:
TR	21 789	:	35.6	2 941	13.5	3 824	17.6
IS	50	1.2	38.0	12	24.0	2	4.0
NO	4 669	7.9	39.2	582	16.3	515	14.4

(1) **Exception to the reference year 2001**
AT: 2000.
(2) ACC: Eurostat estimates.

Source: Eurostat, UOE questionnaire and EU LFS — spring data.

Graduation from tertiary education

Increasing numbers of graduates in EU and Acceding Countries

Though participation rates are a useful proxy for future expectations of the national stocks of HRST, because drop-out rates differ from country-to-country and system-to-system, they should be complemented by data on the actual number of people becoming HRST. Data on graduates measure just that.

In 2001, there were close to 2 million new graduates in the European Union and over 630 thousand in the Acceding Countries – Table 4.3. This compared with just over 1 million new graduates in Japan and over 2.1 million in the United States. Balancing these new graduates against the young population, for every thousand people aged 20-29 in the EU there were on average around 40 new graduates. But this varies from around 70 new graduates per thousand 20-29 year olds in Ireland and France, to between 26 and 27 in Italy and Austria and approximately 34 in Germany.

The Acceding Countries perform better, on average, than the EU. In the Acceding Countries in 2001, there were about 55 new graduates per thousand 20-29 year olds, though being by far the largest Acceding Country, Poland at around 74 new graduates, provides much of the impetus for this higher rate. Latvia and Lithuania also exceeded the EU average, but all other Acceding Countries had fewer new graduates relative to the 20-29 year old population than in the EU.

Table 4.3.

Graduation from tertiary education — in total and selected fields of study by sex in comparison to the population aged 20-29 EU-15, Candidate Countries, Iceland, Norway, Japan and the United States 2001 (1)

	Graduates from tertiary education in 2001						
	Total graduates			In science		In engineering, manufacturing and construction	
	Total	Per 1000 population aged 20-29	% women in total	Total	% of total graduates	Total	% of total graduates
EU-15 (2)	1 963 415	40.4	55.9	218 755	11.1	286 087	14.6
BE	70 202	53.4	56.1	5 704	8.1	7 535	10.7
DK	39 017	55.7	56.3	3 163	8.1	5 293	13.6
DE	296 640	33.7	51.6	26 460	8.9	50 157	17.0
EL	:	:	:	:	:	:	:
ES	277 853	42.6	57.2	29 200	10.5	45 112	16.3
FR	508 189	67.7	55.8	78 074	15.4	76 682	15.1
IE	45 818	70.3	56.0	8 707	19.8	5 331	12.1
IT	202 309	25.8	55.9	15 577	7.7	31 013	15.4
LU	:	:	:	:	:	:	:
NL	81 603	39.8	54.7	4 279	5.2	8 385	10.3
AT	27 099	27.4	51.5	1 840	6.8	5 583	20.7
PT	61 136	37.2	67.1	3 102	5.1	7 155	11.7
FI	36 141	57.7	61.7	2 728	7.5	7 376	20.4
SE	42 741	38.9	58.5	4 329	10.1	9 373	21.9
UK	273 987	36.9	56.6	35 519	13.0	27 066	9.9
ACC (2)	631 073	55.3	63.7	26 758	4.2	55 433	8.8
CZ	43 629	25.9	55.3	4 569	11.0	5 017	12.1
EE	7 600	39.9	65.3	456	6.0	923	12.1
CY	2 813	33.9	65.2	156	5.5	180	6.4
LV	20 308	62.6	55.4	1 032	5.1	1 441	7.1
LT	27 471	51.5	63.5	1 352	4.9	5 673	20.7
HU	57 882	36.1	61.4	1 379	2.4	5 820	10.1
MT	2 003	:	52.0	83	4.1	103	5.1
PL	431 104	74.3	65.9	15 011	4.8	29 831	9.5
SI	11 991	40.0	59.4	437	3.6	1 995	16.6
SK	26 272	29.4	54.2	2 283	8.7	4 450	16.9
BG	47 504	43.7	62.5	1 989	4.2	7 128	15.0
RO	76 230	22.7	54.8	4 333	5.8	14 032	18.9
TR	241 464	:	42.8	19 961	9.6	41 506	20.0
IS	2 066	51.4	62.1	280	13.6	113	5.5
NO	32 092	54.6	58.8	2 675	8.7	2 486	8.1
JP	1 067 878	:	49.4	28 884	2.8	204 502	19.9
US	2 150 954	:	57.0	190 115	8.9	179 276	8.4

NB: EU-15 excludes EL and LU.

(1) **Exceptions to the reference year 2001**
DK, FR, IT, FI, CY, HU and US: 2000.

(2) EU-15 and ACC: Eurostat estimates.

Source: Eurostat, UOE questionnaire and EU LFS — spring data.

Denmark, along with the United Kingdom, saw the highest EU increases in the number of people graduating from tertiary level education between 1998 and 2001 – Figure 4.7. Both had average annual increases of 10% or more, compared to the EU average of 2.5% per year. Germany and Finland were the only EU countries to have experienced a contraction in the number of new graduates. As was the case for enrolment, all of the Acceding Countries experienced a rapid expansion in the number of graduates from higher education.

Compare the proportion of total graduates that were women in Table 4.3. with their participation in Table 4.1. and, with the exception of a few countries, a clear trend emerges. Women accounted for an even higher proportion of all graduates than they did for students. On average, 55.9% of all graduates were women in the EU in 2001, compared with 63.7% in the Acceding Countries, 49.4% in Japan and 57.0% in the United States.

In the EU at least, this appears to be a trend that does not extend to science and engineering related disciplines. Whilst women accounted for 38% of science and 22% of engineering students in the European Union – recall Figure 4.3., they accounted for marginally more science and marginally less engineering graduates at 41% and 21%, respectively – Figure 4.8. In the Acceding Countries, though four in ten science students were women in 2001, nearly half of all science graduates were women (48%). In engineering, women represented 23% of Acceding Country students yet 26% of graduates.

Another notable occurrence is the reduction in the proportion of science and engineering subjects amongst the total when the graduation rates in Table 4.3. are compared with the participation rates in Table 4.1. This implies that the drop-out rates for students from S&E disciplines are higher than for some other tertiary level subjects, corroborated by the proportion of students studying science or engineering related disciplines in 1998 and 1999 – at or close to the beginning of a tertiary education cycle in science or engineering, depending on the country's education system. This phenomenon is especially marked for engineering in the Acceding Countries (8.8% against 14.4%). Nevertheless graduation rates in the EU (11.1% for science and 14.6% for engineering) still compared favourably with both Japan and the United States for science, where they represented 2.8% and 8.9% of all new graduates, and with the United States for engineering (8.4%).

Figure 4.7.

**Annual average growth rates in the number of tertiary education graduates
EU-15, Candidate Countries, Iceland and Norway
1998 to 2001 (1)**

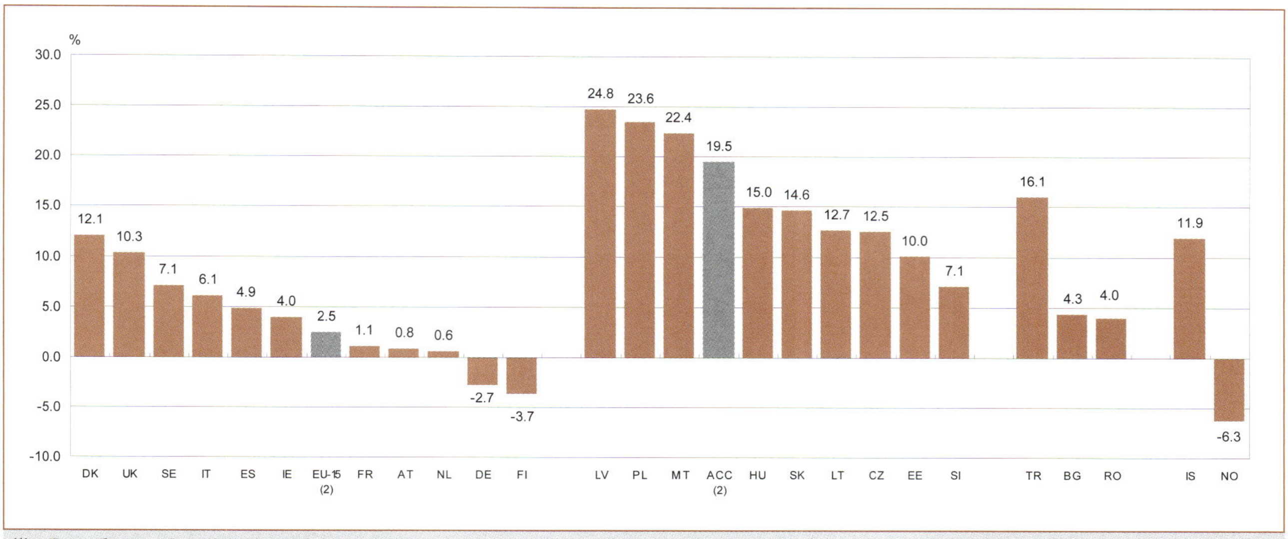

(1) **Exceptions to the reference period 1998-2001**
MT and TR: 1999-2001;
DK, FR, IT and FI: 1998-2000.
(2) EU-15 and ACC: Eurostat estimates.

Source: Eurostat, UOE questionnaire.

Figure 4.8.

**Proportion of female S&E tertiary graduates
EU-15, Candidate Countries, Iceland, Norway, Japan and the United States
2001 (1)**

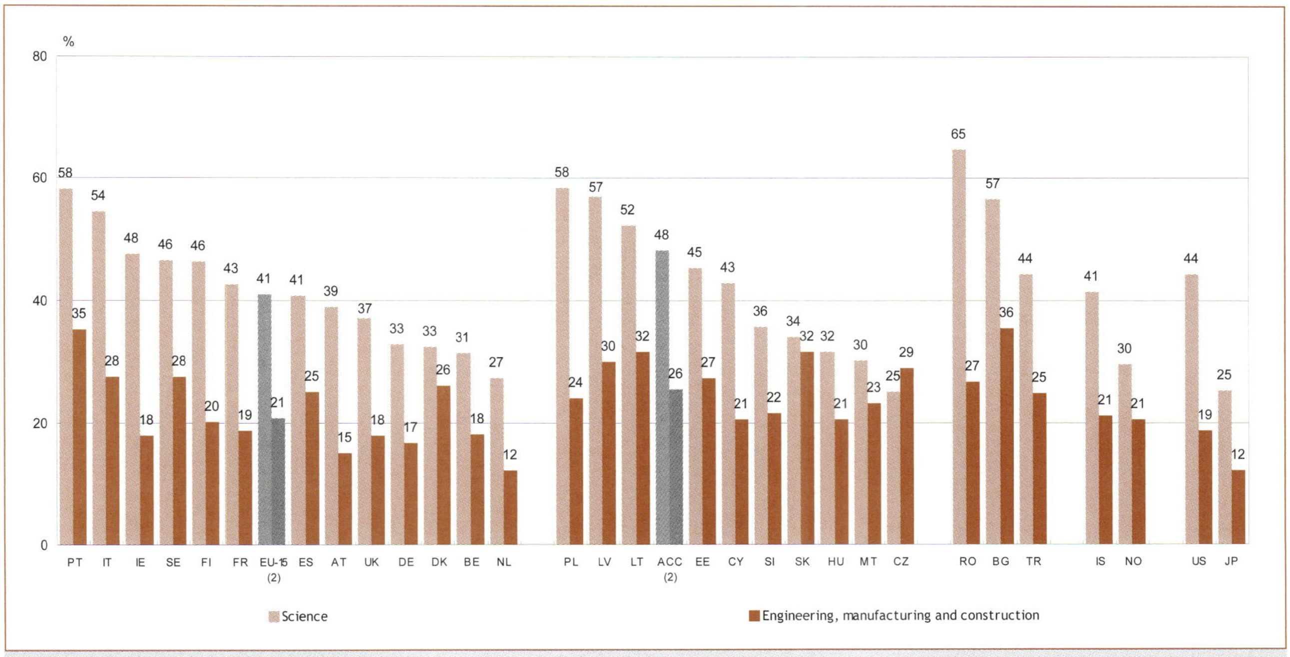

(1) **Exceptions to the reference year 2001**
DK, FR, IT, FI, CY, HU and US: 2000.
(2) EU-15 and ACC: Eurostat estimates.

Source: Eurostat, UOE questionnaire.

PhD graduates

In the EU, 74 908 people obtained their doctoral degree in 2001 compared to just 7 555 in the Acceding Countries, 13 179 in Japan and 44 808 in the United States. Of the EU countries, Germany had by far the highest number of doctorate recipients in 2001, equivalent to a third of all EU PhDs in 2001. Poland led the Acceding Countries, with close to 60% of all Acceding Country doctorates being earned there in 2001.

In engineering, however, PhD graduates in Germany accounted for just under a quarter of reported EU PhD graduates. This was closely followed by the United Kingdom – 22.5% of reported EU PhD graduates.

In the EU Member States, PhD graduates in science accounted for a higher proportion of total PhD graduates in 2001 – Table 4.4. – than did science PhD students among total PhD students – recall Table 4.2. In fact, this is the case in all but one of the EU countries for which data are available.

As far as representation of women is concerned, only in Portugal and Cyprus were there at least as many female PhD graduates as there were male in science. For engineering, the closest to parity was Portugal, where 39% of all PhD graduates in engineering were women. But this fell as low as 12% in Germany for the EU and 20% in Poland for the Acceding countries.

Figure 4.9.

Proportion of female S&E PhD graduates
EU-15, Candidate Countries, Norway, Japan and the United States
2001 (1)

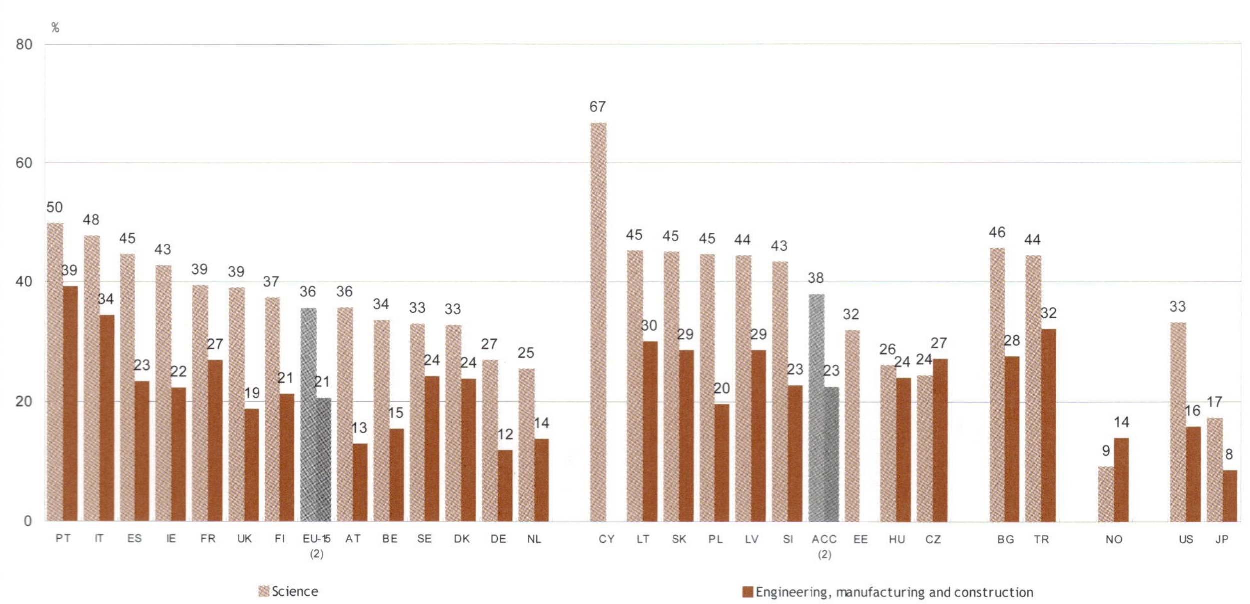

(1) **Exceptions to the reference year 2001**
DK, FR, IT, FI, CY, HU and US: 2000.
(2) EU-15 and ACC: Eurostat estimates.

Source: Eurostat, UOE questionnaire.

Table 4.4.

**Graduation from ISCED level 6 education (PhD) education, in total and selected fields of study by sex in comparison to the population aged 25-29
EU-15, Candidate Countries, Iceland, Norway, Japan and the United States
2001 (1)**

	Graduates at ISCED 6 level (PhDs) in 2001						
	Total graduates at ISCED 6 level (PhDs)			In science		In engineering, manufacturing and construction	
	Total	Per 1000 population aged 25-29	% women in total	Total	% of total ISCED 6 graduates	Total	% of total ISCED 6 graduates
EU-15 (2)	74 908	2.9	39.6	23 149	30.9	9 754	13.0
BE	1 317	1.9	31.9	521	39.6	169	12.8
DK	795	2.0	37.4	190	23.9	207	26.0
DE	24 796	5.5	35.3	6 831	27.5	2 333	9.4
EL	:	:	:	:	:	:	:
ES	6 453	1.9	42.9	1 842	29.4	538	8.6
FR	10 404	2.6	42.7	1 761	48.0	956	9.2
IE	572	1.8	44.4	293	51.3	63	11.0
IT	4 044	0.9	50.8	821	20.3	808	20.0
LU	:	:	:	:	:	:	:
NL	2 533	2.3	31.5	530	20.9	390	15.4
AT	1 871	3.4	37.1	405	21.7	400	21.4
PT	2 791	3.4	50.7	434	15.5	468	16.8
FI	1 797	5.8	45.8	345	19.2	321	17.9
SE	3 388	5.8	39.2	746	22.0	911	26.9
UK	14 147	3.6	39.5	5 202	36.8	2 190	15.5
ACC (2)	7 555	1.3	41.1	1 472	19.5	1 196	15.8
CZ	1 066	1.2	34.7	349	32.7	207	19.4
EE	149	1.5	51.7	22	14.8	9	6.0
CY	13	0.3	76.9	3	23.1	:	:
LV	37	0.2	48.6	18	48.6	7	18.9
LT	261	0.9	52.5	42	16.1	60	23.0
HU	793	1.0	38.0	142	17.9	50	6.3
MT	6	:	:	:	:	1	16.7
PL	4 400	1.6	41.6	709	16.1	679	15.4
SI	298	2.0	49.0	76	25.5	57	19.1
SK	532	1.2	39.8	111	20.9	126	23.7
BG	376	0.7	42.0	68	18.1	58	15.4
RO	:	:	:	:	:	:	:
TR	1 985	:	38.4	320	16.1	320	16.1
IS	3	0.2	100.0	:	:	:	:
NO	768	2.4	34.4	11	2.1	79	15.4
JP	13 179	:	22.8	2 070	15.8	3 048	23.2
US	44 808	:	44.1	10 768	24.1	5 519	12.3

NB: EU-15 excludes EL and LU.

(1) **Exceptions to the reference year 2001**
DK, FR, IT, FI, CY, HU and US: 2000.

(2) EU-15 and ACC: Eurostat estimates.

Source: Eurostat, UOE questionnaire and EU LFS — spring data.

4.3. Stocks of human resources in science and technology

On the supply side, we have seen that, in general, the number of participants in tertiary education has been increasing, as has the number of graduates. On the demand side, this section now looks at the labour markets in the various EU Member States and Acceding Countries themselves. The measurement of stocks of HRST and of its various sub-categories provides broad indicators on the state of the labour markets in the different European countries.

HRST stocks at the national level

Table 4.5. shows the stocks of HRST in 2002 as well as growth in the number of S&T jobs over time. Germany, France and the United Kingdom in 2002 had the highest numbers of HRST, together accounting for close to 57% of the EU's 62 million HRST. Compared to the EU average of just under half all HRST being female, both Germany and the United Kingdom fell short. Parity or better was reached in Denmark, Ireland, Portugal, Finland and Sweden. In the Acceding Countries, however, HRST were far more likely to be women (57.2%), and as many as six in ten S&T jobs – HRSTO – in the Acceding Countries were carried out by women.

The highest growth rates in the number of people working in S&T occupations can be found in Spain and Italy in the EU, and Cyprus and Slovenia in the Acceding Countries. Though equal in the Acceding Countries, growth in the number of S&T jobs was higher for women than for men in the EU overall (3.5% against 2.2%). With those rates, and other things being equal, parity between the number of women and men working in S&T jobs in the enlarged European Union (i.e. EU-25) should be reached by 2007/2008.

The largest differences between growth in the number of S&T jobs for men and for women tend to occur in the countries where the disparity between existing stocks of the two is the greatest. The proportion of women HRSTO in Luxembourg for example is the lowest in the EU, yet the difference between growth rates for men and for women, one of the highest. A similar effect can be seen in Cyprus. Where men are under-represented, on the other hand, the opposite growth rates tend to prevail – Finland and Estonia. This further indicates a tendency towards parity between the number of male and female S&T workers.

Although Germany has by far the highest number of HRST overall, it also has the largest population. However, Figure 4.10. shows that, in 2002, Germany lay just above the EU average when the number of 25-64 year olds with tertiary education were measured against the entire same age group (21%). Finland headed the EU countries with almost a third of 25-64 year olds being tertiary educated, followed by Belgium and Denmark. Italy and Portugal were far below the EU average and also that for the Acceding Countries.

Figure 4.10.

Proportion of 25-64 year olds with tertiary education
EU-15, Candidate Countries, Iceland, Norway and Switzerland
2002 (1)

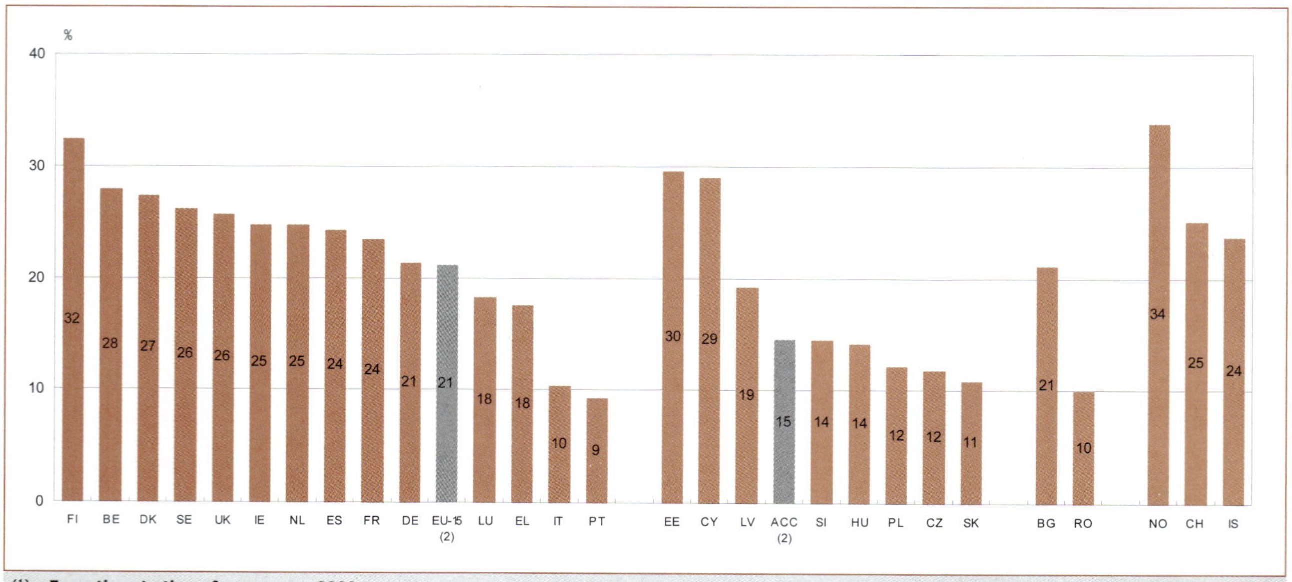

(1) **Exceptions to the reference year 2002**
UK: 2000;
IS: 2001.
No data available for AT and LT.
(2) EU-15 and ACC: Eurostat estimates.

Source: Eurostat, EU LFS — spring data.

Table 4.5.

**Stocks of 25-64 year old HRST by country and sex in 2002 (1) and growth in S&T occupations between 1998 and 2002 (2)
EU-15, Candidate Countries, Iceland, Norway and Switzerland**

	HRST		HRSTE		HRSTO		HRSTC		Annual average growth rate of HRSTO 1998-2002	
	1000s	% Women	1000s	% Women	1000s	% Women	1000s	% Women	% Men	% Women
EU-15 (3)	61 986	47.0	43 419	47.5	42 806	47.3	24 239	48.5	2.2	3.5
BE	1 885	48.6	1 539	50.7	1 132	48.1	786	52.1	1.2	2.5
DK	1 168	50.8	809	54.0	906	51.9	547	57.4	2.9	5.2
DE	15 702	45.5	9 814	39.5	11 287	49.1	5 399	42.1	0.7	1.9
EL	1 126	47.5	967	48.2	725	47.6	566	48.8	-0.9	1.6
ES	6 243	47.4	5 411	49.1	3 441	45.5	2 609	48.5	5.5	7.9
FR	9 531	48.8	7 160	52.4	6 469	46.3	4 099	51.3	2.4	3.2
IE	585	51.7	489	52.1	339	51.8	244	52.7	2.9	6.2
IT	6 911	46.6	3 345	49.6	5 737	45.2	2 170	47.6	5.0	7.3
LU	70	43.3	45	41.3	57	43.1	33	40.3	-0.6	3.5
NL	3 428	46.3	2 219	44.3	2 517	47.5	1 308	45.2	0.1	3.6
AT	:	:	:	:	873	48.3	:	:	1.2	3.8
PT	794	52.8	516	62.0	663	51.9	385	63.6	3.1	4.5
FI	1 134	53.4	916	55.0	729	52.6	510	55.2	4.3	-0.7
SE	1 901	50.3	1 238	56.1	1 544	50.0	881	58.0	3.6	4.7
UK	9 895	45.0	8 188	47.1	6 304	46.0	4 141	47.4	2.9	4.1
ACC (3)	9 231	57.2	5 711	55.1	6 747	59.7	3 228	58.8	1.1	1.0
CZ	1 504	50.0	679	42.5	1 274	52.0	449	44.6	3.1	1.2
EE	258	64.0	212	63.3	142	66.7	95	66.5	7.1	0.0
CY	123	47.5	105	48.4	75	45.6	56	46.6	5.4	14.7
LV	355	61.0	239	59.5	227	66.6	111	69.4	1.0	1.7
LT	:	:	:	:	315	70.0	:	:	3.2	2.1
HU	1 171	57.3	763	52.8	849	60.9	442	56.5	0.7	1.8
MT	:	:	:	:	:	:	:	:	:	:
PL	4 037	58.7	2 450	57.1	3 074	61.0	1 488	60.9	-0.2	-0.1
SI	294	54.2	161	54.9	235	55.9	102	59.3	4.2	7.0
SK	660	58.7	305	51.6	557	61.0	202	54.2	0.9	2.8
BG	1 103	58.8	908	60.5	661	60.9	465	65.1	0.6	-1.2
RO	1 944	53.2	1 182	46.6	1 526	56.6	764	49.8	-0.2	1.1
TR	:	:	:	:	:	:	:	:	:	:
IS	51	52.6	33	50.0	41	55.3	22	53.7	4.1	8.9
NO	1 043	48.4	815	50.5	739	48.8	511	52.3	5.5	5.3
CH	1 652	41.5	1 019	33.0	1 248	44.0	614	32.6	0.0	3.4

(1) **Exceptions to the reference year 2002**
HRST, HRSTE and HRSTC — IS: 2001;
HRST, HRSTO and HRSTC — UK: 2000.
(2) **Exceptions to the reference period 1998-2002**
IE, CY and SK: 1999-2002;
UK: 1998-2000;
BG: 2000-2002;
IS: 1998-2001.
(3) EU-15 and ACC: Eurostat estimates.

Source: Eurostat, EU LFS — spring data.

Employment rates — evaluating the level and quality of employment

Looking at employment overall, it is possible to distinguish among those people that are working in a S&T occupation who have a third level education — HRSTC, those working in a S&T occupation without a third level education — HRSTO excluding HRSTC — and finally those employed, but not working in a S&T occupation. This way, an insight into the types of jobs as well as the qualification level of the people carrying them out can be obtained.

Sweden had the highest proportion of its working population in S&T occupations, having obtained a third level education in 2002 (around 910 000 or around 21%). If you include those working in S&T without a third level education, then this rose to more than 1.6 million (38% of the total workforce). The Netherlands, Denmark and Germany followed, with around 35%, all far above the EU average of 28.6% — Figure 4.11.

The average in the Acceding Countries, meanwhile, was not so different to the EU. There, 25.4% of all jobs in 2002 were in S&T. In the Czech Republic and the Slovak Republic, where the highest Acceding Country rates can be found, the proportion of the workforce engaged in S&T occupations was on a par with France at 29%.

Figure 4.11. **Employment distribution of 25-64 year olds, in thousands and proportion of people working in S&T EU-15, Candidate Countries, Iceland, Norway and Switzerland 2002 (1)**

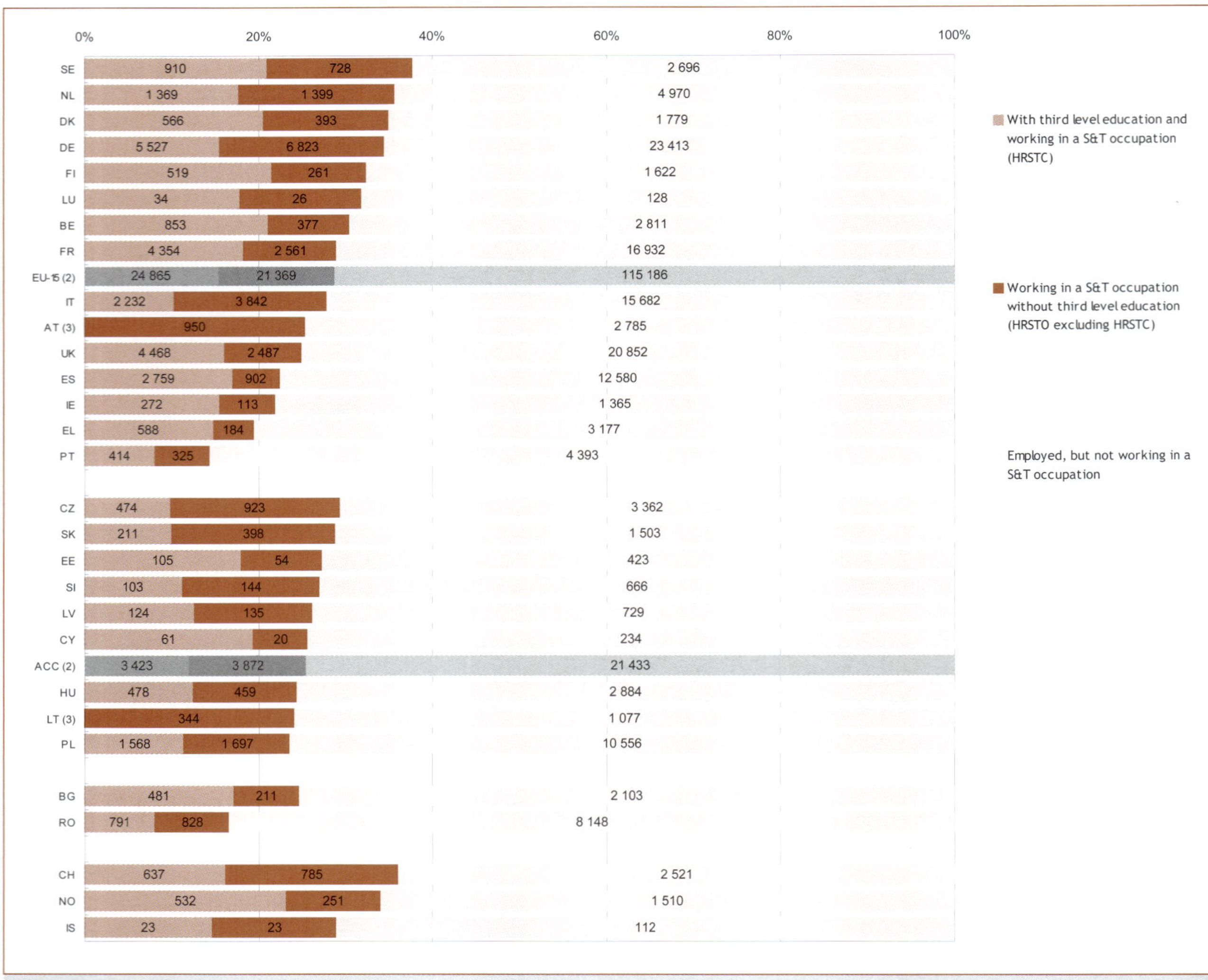

(1) **Exceptions to the reference year 2002**
UK: 2000;
IS: 2001.
(2) EU-15 and ACC: Eurostat estimates.
(3) No HRSTC data available for AT and LT.
Therefore people working in a S&T occupation may or may not have a third level education.

Source: Eurostat, EU LFS — spring data.

The ageing workforce

If the previous data on HRSTO in Figure 4.11. are complemented with those in Figure 4.12., then the increased proportion of HRSTO amongst the young in the Acceding Countries bears out the recent increases in the number of Acceding Country graduates in Table 4.3. Around 15% of all S&T workers — HRSTO — in the Acceding Countries were aged 25-29. Add the strong growth in the number of students in tertiary education that will enter the HRST stocks once they graduate as well as the fact that, at the present time, the population is generally younger in the Acceding Countries — 18% of the Acceding Country population is under 25 — and they should continue to catch up with the better performing EU countries in the medium term. In the longer term, recently falling birth rates in the Acceding Countries may be exacerbated by the retirement of the high proportion of Acceding Country residents in their forties — see population in Figure 4.12.

Figure 4.12. **Age distribution of employed S&T workers — HRSTO — and the total population EU-15 and Acceding Countries 2002 (1)**

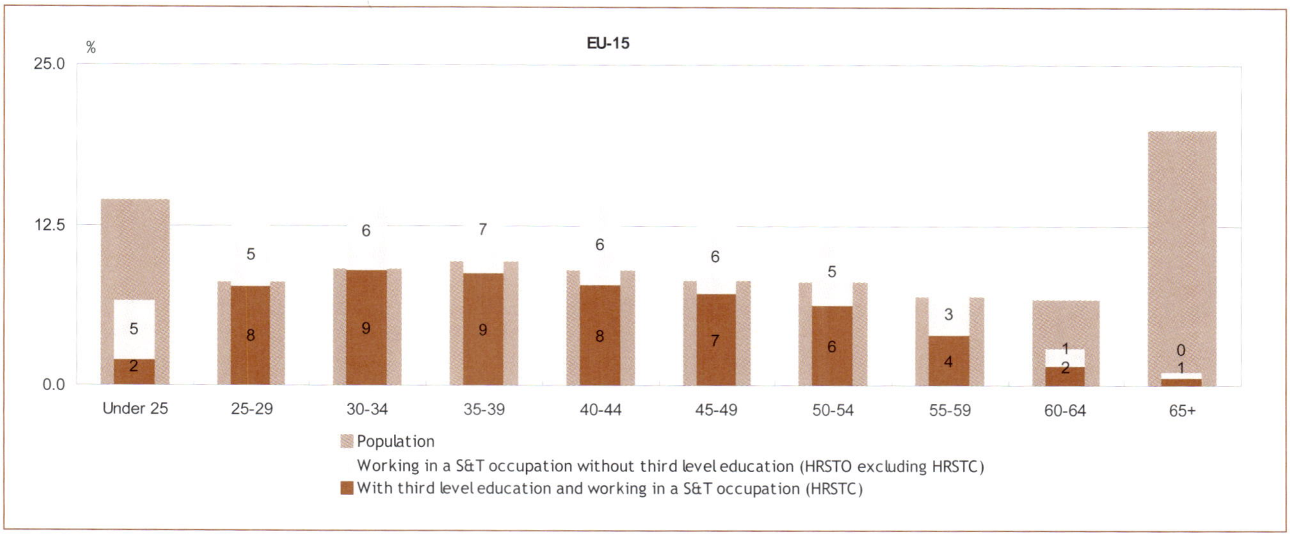

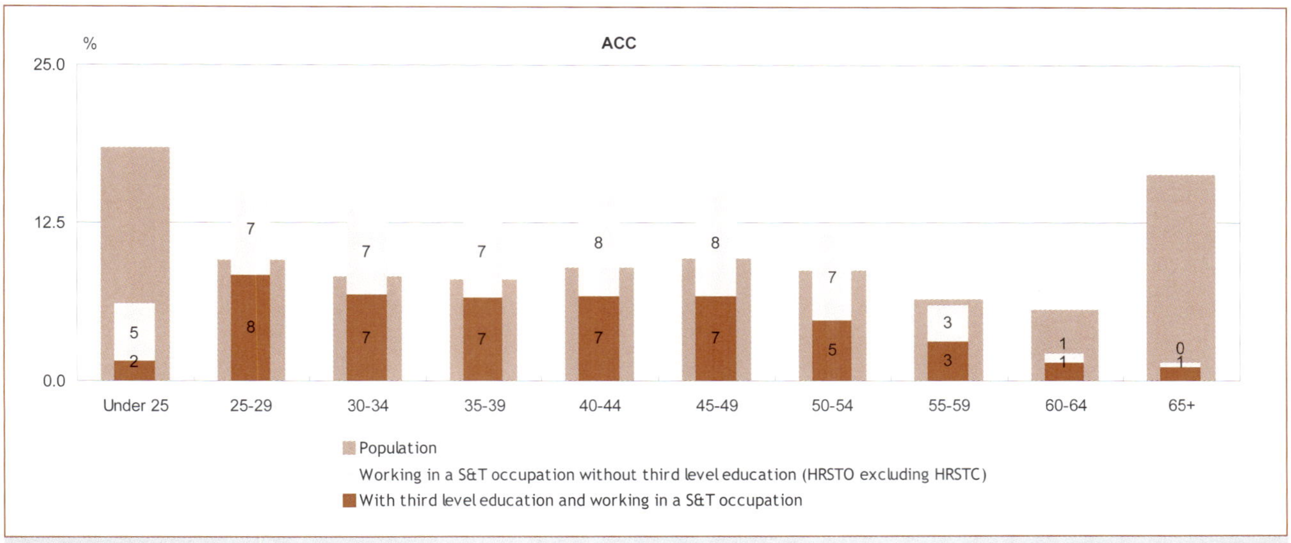

(1) EU-15 and ACC: Eurostat estimates.

Source: Eurostat, EU LFS — spring data.

Scientists and engineers

Scientists and engineers are predominantly male occupations

Figure 4.13. shows the gender distribution of scientists and engineers — S&E — in 2002, measured as a percentage of the total labour force. In all EU Member States except Ireland, being a scientist or engineer was a predominantly male occupation — although in Ireland, scientists and engineers were just as likely to be women than men. Indeed in most countries, differences were high, with three Member States in 2002 showing a ratio of close to four male to every female scientist or engineer: Germany, Luxembourg and France.

Being a scientist or engineer was, on average, less common in the Acceding Countries than in the EU. But whilst this may be true, the gender balance between those people in the labour force working in S&E was far more even. In both Latvia and Lithuania, scientists and engineers were more likely to be women than men, where in total, close to 4% of the labour force were scientists or engineers.

Figure 4.13. **Distribution of 25-64 year old scientists and engineers by sex as a % of the total labour force EU-15, Candidate Countries, Iceland, Norway and Switzerland 2002 (1)**

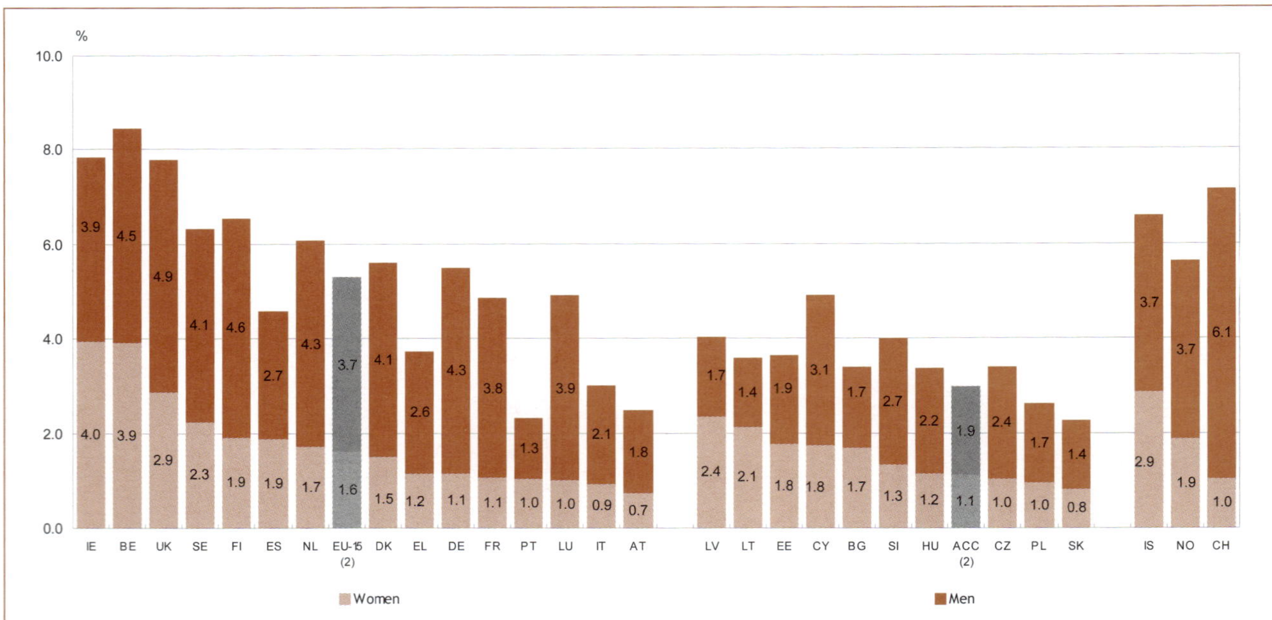

(1) **Exceptions to the reference year 2002**
UK: 2000;
IS: 2001.
(2) EU-15 and ACC: Eurostat estimates.

Source: Eurostat, EU LFS — spring data.

Esteem for different professions

Figure 4.13. showed the proportion of the labour force working as either a scientist or engineer in 2002. However, an important indicator for future growth in these professions is the opinion that society has for these occupations. Table 4.6. shows the esteem people hold for various professions in the different EU Member States. This shows science and engineering professions, behind doctors, to be those for which people have the highest esteem: 44.9% of people in the EU place scientists as second highest and 29.8% engineers as third highest.

Table 4.6.

Esteem for different professions
in % of answers

Question: for which of the following professions do you have the most esteem?

	EU-15	BE	DK	DE	EL	ES	FR	IE	IT	LU	NL	AT	PT	FI	SE	UK
Doctors	71.1	74.3	58.9	64.4	68.0	68.0	80.4	69.6	67.4	79.2	72.2	65.2	76.5	76.0	73.9	78.0
Scientists	44.9	48.5	50.1	42.7	53.3	47.4	47.9	22.9	46.4	50.1	50.0	36.2	35.2	43.5	54.8	40.9
Engineers	29.8	31.5	28.7	26.6	24.7	32.1	33.8	24.3	27.1	31.9	29.2	16.5	26.4	27.5	24.5	36.3
Judges	27.6	21.3	41.9	35.5	26.0	20.9	20.0	24.0	23.3	32.5	39.1	29.0	30.4	26.3	37.4	27.2
Sportsmen	23.4	30.5	14.7	16.8	49.1	32.8	26.3	35.0	19.3	22.5	27.5	23.1	22.3	17.1	12.9	23.3
Artists	23.1	32.2	19.2	16.4	31.8	25.8	30.3	13.4	29.8	26.4	29.6	13.7	24.9	25.6	17.5	14.8
Lawyers	18.1	17.4	21.3	21.1	17.5	15.2	15.4	16.2	12.5	20.3	24.7	15.6	15.5	14.0	20.3	22.8
Journalists	13.6	20.3	8.8	8.6	24.4	26.7	17.6	14.1	12.3	26.8	15.9	8.1	25.8	10.0	9.3	5.0
Businessmen	13.5	17.8	11.9	9.0	14.5	16.0	10.6	18.4	18.1	17.1	13.7	16.0	15.6	18.6	11.2	14.6
Politicians	6.6	8.7	13.1	7.8	5.8	6.2	3.2	6.1	4.5	16.8	14.9	8.7	5.9	7.1	9.8	6.3
None of them	6.9	4.7	7.9	8.9	6.5	8.0	5.6	6.2	6.7	3.6	7.6	9.1	4.8	4.0	6.9	5.1
Don't know	3.0	2.6	3.0	3.5	0.4	4.2	1.5	5.5	2.5	2.8	3.4	3.4	3.3	2.0	2.7	3.6

***Source**: Third European Report on Science and Technology Indicators*, Directorate-General for Research, 2003, p. 200.
***Data**:* Eurobarometer 55.2, Table 26.

HRST intensity by sector of activity

High technology most knowledge-intensive manufacturing industry

HRST intensity can be defined as the number of employed people with a third level education — employed HRSTE — as a ratio of total employment. In turn, this can be seen as a proxy for the knowledge intensity — the proportion of highly qualified people — in each sector of economic activity.

Table 4.7. shows the list of grouped sectors of activity — classified according to NACE Rev.1.1 — used for presenting the results. The grouping for manufacturing sectors respects the breakdowns made by the OECD for their measurement of technology sectors according to R&D intensity, though the OECD definition is based on 3 digit level NACE, whereas the data are only available at the 2 digit level, whilst services are broken down according to the Eurostat definitions for knowledge intensity in the service sectors.

Table 4.7. **List of NACE sector groups for measurement of knowledge intensity**

Description	NACE Rev 1.1 codes
Agriculture, hunting, forestry, fishing, mining and quarrying	01 to 14
Utilities and construction	40, 41 and 45
Low-technology	15 to 22 and 37
Medium low technology	23, 25 to 28 and 36
Medium high tech manufacturing	24, 29, 31, 34 and 35
High tech manufacturing	30, 32 and 33
Knowledge-intensive high technology services	64, 72 and 73
Knowledge-intensive market services (excl. financial intermediation and high-tech services)	61, 62, 70, 71 and 74
Knowledge-intensive financial services	65, 66 and 67
Other knowledge-intensive services	80, 85 and 92
Less-knowledge-intensive market services	50, 51, 52, 55, 60 and 63
Other less-knowledge-intensive services	75, 90, 91, 93, 95 and 99

Source: Eurostat.

High technology manufacturing was the most knowledge-intensive of the manufacturing industries in the EU in 2002 where around a third of all employed people were tertiary educated — Table 4.8. Finland displayed the highest EU rate at 54.9% and Italy the lowest (14.8%). Meanwhile, utilities and construction (25.4%) was slightly more knowledge-intensive than medium high technology manufacturing sectors in the EU in the same year (23.4%).

In the Acceding Countries, high technology manufacturing employment stands at 14.3%, far below its EU counterpart.

Table 4.8. **Knowledge intensity of employed 25-64 year olds in agriculture, manufacturing and utilities EU-15, Candidate Countries, Iceland, Norway and Switzerland 2002**

	% of employed 25-64 year olds that are HRSTE					
	Agriculture, hunting, forestry, fishing, mining and quarrying	Utilities and construction	High tech manufacturing	Medium high tech manufacturing	Medium low technology manufacturing	Low technology manufacturing
EU-15 (1)	**9.1 s**	**25.4 s**	**32.9 s**	**23.4 s**	**13.0 s**	**12.8 s**
BE	17.5	23.8	44.9	29.5	20.8	20.7
DK	9.5 u	23.9	34.5	22.8	12.0	13.4
DE	19.3	27.7	31.9	25.3	15.1	16.4
EL	1.2	33.8	34.1 u	18.8	11.1	9.4
ES	7.0	32.2	49.4	36.7	21.9	16.2
FR	8.6	21.3	39.0	23.0	12.5	14.4
IE	5.8	27.0	39.7	31.2	16.7	17.3
IT	2.9 p	13.9 p	14.8 p	9.1 p	3.5 p	3.9 p
LU	: u	20.7 u	: u	: u	13.3 u	: u
NL	8.7	17.5	33.3	18.3	11.5	12.9
AT	:	:	:	:	:	:
PT	: u	19.8	: u	7.5 u	4.4 u	2.4 u
FI	16.5	26.6	54.9	31.9	21.9	23.6
SE	9.2 p	13.1 p	32.2 p	17.0 p	6.0 p	9.8 p
UK	21.0	27.5	35.5	27.6	16.5	16.5
ACC (1)	**10.3 s**	**22.1 s**	**14.3 s**	**11.6 s**	**7.3 s**	**10.1 s**
CZ	6.7	15.1	12.1	10.4	5.8	5.2
EE	19.4 u	36.5 u	: u	21.8 u	: u	18.9 u
CY	4.2 u	25.9	: u	36.1 u	7.9 u	15.6
LV	7.1	29.3	: u	: u	9.8	13.7
LT	:	:	:	:	:	:
HU	7.4	24.1	12.0	9.6	6.4	5.8
MT	:	:	:	:	:	:
PL	:	:	:	:	:	:
SI	4.7 u	15.6 u	19.0 u	10.3 u	7.1 u	7.3 u
SK	3.1	16.6	13.4 u	6.1	5.9	3.7
BG	5.9	22.1	: u	19.8	14.5	11.4
RO	2.0	23.2	17.8	12.9	8.8	5.6
TR	:	:	:	:	:	:
IS	: u	: u	: u	: u	: u	: u
NO	23.5 p	27.2 p	71.9 p	26.9 p	14.6 p	17.8 p
CH	14.5	29.3	33.7	33.6	15.2	19.1

(1) EU-15 and ACC: Eurostat estimates.

Source: Eurostat, EU LFS — spring data.

Despite country level differences, however, at the aggregate level, knowledge intensive employment in services follows a similar structure in the EU and the Acceding Countries — see Table 4.9. Other knowledge intensive services, which include 'Education' and 'Heath and social work', have the highest proportion of tertiary educated employed people in both the EU and the Acceding Countries.

Table 4.9.

Knowledge intensity of employed 25-64 year olds in services EU-15, Candidate Countries, Iceland, Norway and Switzerland 2002

	% of employed 25-64 year olds that are HRSTE					
	Knowledge-intensive financial services	Knowledge-intensive high technology services	Knowledge-intensive market services (excl. financial intermediation and high tech services)	Other knowledge-intensive services	Less-knowledge-intensive market services	Other less-knowledge-intensive services
EU-15 (1)	32.8 s	38.6 s	37.3 s	47.9 s	12.8 s	26.5 s
BE	58.4	45.5	53.2	61.0	17.6	28.1
DK	25.2	36.1	38.3	49.5	13.9	34.1
DE	24.5	34.0	32.4	41.3	12.6	31.0
EL	37.5	37.4	52.9	65.5	10.6	31.2
ES	52.2	59.6	45.8	66.5	18.2	32.2
FR	42.9	44.9	38.8	49.4	16.9	22.1
IE	49.4	49.5	45.9	54.6	15.2	28.2
IT	21.4 p	21.4 p	30.5 p	37.1 p	5.2 p	13.5 p
LU	30.0	26.1 u	33.7	46.4	6.9	23.4
NL	34.4	45.3	38.8	45.2	12.3	33.8
AT	:	:	:	:	:	:
PT	27.5	28.7	24.7	41.5	3.9	10.7
FI	60.1	51.9	44.3	50.8	25.2	49.3
SE	29.9 p	41.6 p	29.1 p	46.2 p	11.8 p	40.8 p
UK	32.5	38.3	40.0	49.8	13.2	32.0
ACC (1)	37.7 s	33.9 s	35.3 s	43.5 s	14.1 s	26.6 s
CZ	32.6	28.8	27.1	33.2	6.3	18.6
EE	: u	57.1 u	46.8	55.8	27.5	37.7
CY	57.8	67.5	57.9	74.0	24.1	30.8
LV	49.6 u	47.6 u	29.5	39.4	19.9	29.0
LT	:	:	:	:	:	:
HU	32.0	27.9	32.9	43.7	10.3	24.0
MT	:	:	:	:	:	:
PL	:	:	:	:	:	:
SI	31.6 u	28.3 u	28.7 u	41.9	9.3	32.8
SK	35.0	34.1	32.2	33.6	7.3	20.0
BG	52.3	44.1	51.3	65.2	19.9	35.8
RO	42.6	39.0	34.0	33.6	11.3	26.0
TR	:	:	:	:	:	:
IS	30.9	45.6	35.4	42.5	14.3	30.2
NO	48.4 p	50.3 p	43.1 p	53.7 p	18.1 p	48.3 p
CH	32.9	33.9	40.5	37.5	16.3	32.8

(1) EU-15 and ACC: Eurostat estimates.

Source: Eurostat, EU LFS — spring data.

Unemployment

Completing tertiary education significantly reduces the risk of unemployment, especially in Acceding Countries

This section provides an insight into the unemployment rates for those that have a third level education as well as how these rates compare to those people that do not have the same level of formal qualification.

As Figure 4.14. shows, unemployment rates in 2002 were similar for the tertiary educated in the EU and the Acceding Countries — HRSTU: both stood at 3%. This put tertiary educated unemployment in the EU at just over 1.6 million people and around 230 000 in the Acceding Countries. Deviation from the average was small in both groups of countries, with the highest unemployment rate evident in Spain in the EU and Bulgaria in the Acceding Countries (6%).

Finding and retaining a job when you do not possess third level education, however, is more difficult. In 2002, there were over 11 million sub-tertiary educated unemployed in the EU and 4.7 million in the Acceding Countries. This translates into an unemployment rate of 9% and 17%, respectively.

The starkest contrasts in unemployment rates between those that have a third level education and those that do not can be found in Poland and the Slovak Republic. Indeed, it is a phenomenon that is generally far more accentuated in the Acceding Countries than in the EU, where the most notable cases were Finland, Italy and Germany.

Figure 4.14.

**Unemployment rates for tertiary and non-tertiary educated people aged 25-64
EU-15, Candidate Countries, Iceland, Norway and Switzerland
2002 (1)**

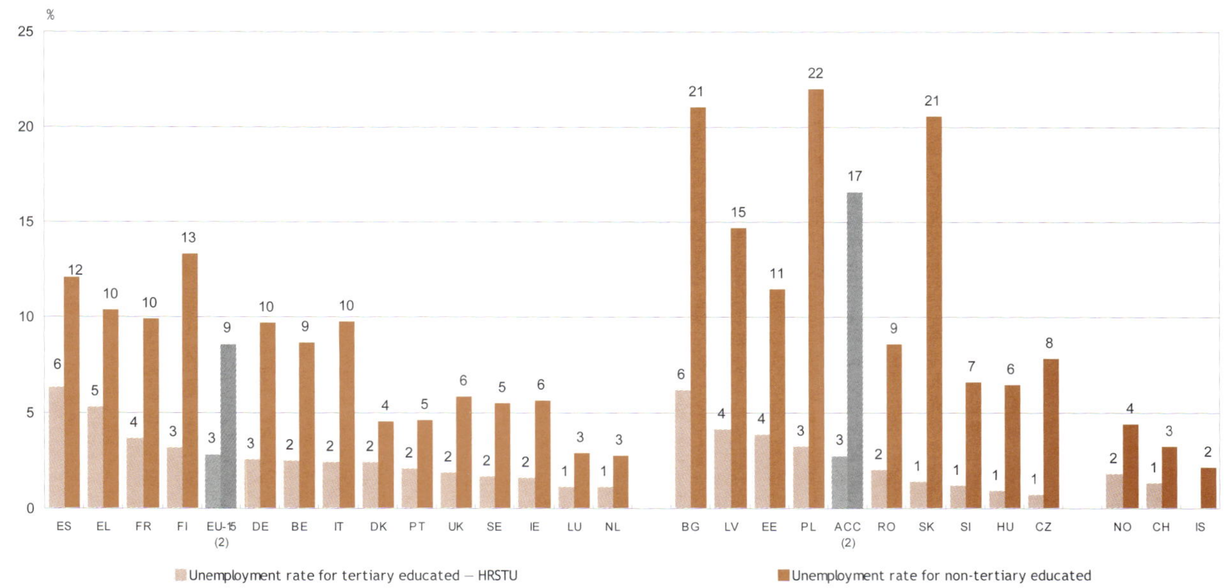

(1) **Exception to the reference year 2002**
IS: 2001.
Reliable data not available for
AT, CY and LT;
IS for HRSTU.
HRSTU data for LU and SI should be treated with caution.
(2) EU-15 and ACC: Eurostat estimates.

Source: Eurostat, EU LFS — spring data.

HRST stocks at the regional level

When analysing EU LFS data at the regional level, particular attention needs to be paid to their reliability. This is because the size of the samples, for which the aim is to provide a representative estimate of the population of that region, can become small and be prone to sampling error. This is especially true when the data are also disaggregated by sector of activity. For this reason, data by sector of activity are presented at the NUTS 1 regional level only, whilst totals are presented at the NUTS 2 level (5). In any case, a strict adherence to the guidelines provided by the *European Union Labour Force Survey vis-à-vis* the minimum levels at which data can be considered reliable has been employed.

In most cases, data are well above the minimum sample size guidelines provided by the Employment Unit of Eurostat for using the *European Union Labour Force Survey*. Data are flagged accordingly when this is not the case.

The top 30 HRST regions in the European Union

As a reminder, HRSTE are defined as having successfully completed a tertiary education, HRSTO as being employed in a S&T occupation, HRST as either of the above and HRSTC as both.

Ranked according to the number of people in the labour force who are HRST, Stockholm is the leading region, where compared to the labour force over half of all residents either had third level education or worked in S&T in 2002 (53.1%). Over half of residents in Uusimaa (FI) were HRST compared to the labour force (52.4%). Germany, however, had the most regions in the top 30, with seven. The Netherlands had six, Sweden and the United Kingdom four, Belgium three and Finland and Spain two each — Table 4.10.

A high proportion of the active population in S&T does not necessarily mean that these people have third level education. As the HRSTC column shows, there can be a wide variety in the level of formal education of S&T workers. Take Hamburg and Berlin. Both had close to 40% of the labour force working in S&T jobs in 2002 — HRSTO. Yet in Hamburg, only 17% of the labour force worked in S&T and had a tertiary education compared to over 22% in Berlin — HRSTC.

(5) The regional data presented in this publication are broken down according to the ***Nomenclature of Territorial Units for Statistics*** — NUTS — classification, 1998 version.

Table 4.10.

The top 30 regions in the European Union ranked according to the proportion of the labour force in S&T occupations 2002 (1)

Ranking	Country	Region – NUTS 2	HRST	% of labour force	HRSTE	% of labour force	HRSTO	% of labour force	HRSTC	% of labour force
		EU-15 (2)	61 986	40.5	43 419	28.4	42 806	28.0	24 239	15.9
1	SE	Stockholm	539 p	53.1 p	350 p	39.6 p	439	49.7	250 p	28.3 p
2	FI	Uusimaa (Suuralue)	426 p	52.4 p	329	48.0	282 p	41.1 p	185 p	27.0 p
3	NL	Utrecht	306	48.1	215	42.7	216	42.9	124	24.7
4	BE	Brabant Wallon	86	45.4	75	55.2	49	36.0	39	28.3
5	FR	Île de France	2 723	44.7	2 133	43.1	1 780	35.9	1 190	24.0
6	UK	Inner London	701	44.4	642	52.6	410	34.9	300	25.5
7	NL	Noord-Holland	654	44.4	448	39.0	470	40.9	265	23.0
8	DE	Berlin	839	42.2	583	38.6	594	39.3	338	22.4
9	UK	Berks., Bucks. & Oxfords.	481	41.0	394	39.0	302	30.3	215	21.6
10	BE	Région Bruxelles-capitale	212	40.6	190	52.7	102	28.3	80	22.2
11	DE	Oberbayern	965	40.4	592	31.7	714	38.2	341	18.2
12	SE	Västsverige	369 p	39.9 p	242 p	30.6 p	301	38.0	174 p	22.0 p
13	UK	Surrey, East and West Sussex	542	39.8	429	37.6	347	31.0	217	19.3
14	ES	Comunidad de Madrid	1 183	39.7	1 012	45.8	728	33.0	557	25.2
15	NL	Zuid-Holland	748	39.5	480	32.9	558	38.3	290	19.9
16	DK	Denmark	1 168	39.5	809	33.3	906	37.3	547	22.5
17	SE	Sydsverige	264 p	38.9 p	178 p	31.8 p	210	37.7	124 p	22.2 p
18	NL	Flevoland	74	38.9	44	29.4	55	37.1	25	16.8
19	BE	Vlaams Brabant	216	38.8	173	40.9	126	29.8	83	19.7
20	DE	Darmstadt	838	38.6	509	30.7	618	37.3	289	17.4
21	DE	Dresden	353	38.2	283	38.3	225	30.4	154	20.9
22	ES	Pais Vasco	453	37.9	409	46.6	225	25.6	180	20.5
23	NL	Groningen	118	37.9	75	31.8	86	36.4	44	18.6
24	DE	Hamburg	388	37.9	217	28.4	301	39.5	130	17.0
25	NL	Gelderland	402	37.5	254	30.6	302	36.3	155	18.6
26	SE	Östra Mellansverige	295 p	37.4 p	190 p	29.0 p	239	36.4	134 p	20.4 p
27	UK	Gloucesters., Wilts. and North Somerset	431	37.2	337	34.7	277	28.3	194	19.8
28	DE	Tübingen	360	37.2	231	30.4	251	32.9	122	16.1
29	DE	Karlsruhe	550	36.7	343	29.9	403	35.1	197	17.1
30	FI	Etelä-Suomi	362 p	36.6 p	299	38.1	231 p	29.4 p	169 p	21.4 p

(1) **Exceptions to the reference year 2002**
UK: 2000;
IS: 2001.
(2) EU-15: Eurostat estimates.

Source: Eurostat, EU LFS — spring data.

Sectoral differences in the top 30 HRSTE regions in the European Union

Regional HRST data can also be broken down by sector of activity – though at the NUTS 1 regional level instead of NUTS 2. Tables 4.11. and 4.12. show, for each region, the proportion of employed people with a third level education. Région Bruxelles-capitale (BE) had the highest proportion of total employment with tertiary education for the manufacturing industry as a whole (37.3%). In the high and medium high technology sector, as many as 6 in every ten employed people also had a third level education in Région Bruxelles-capitale (BE, 59.3%), far ahead of second place Comunidad de Madrid (ES) at 47.5% and third place London (UK, 43.5%).

Table 4.11.

The proportion of S&T in manufacturing industries the top 30 regions in % 2002 (1)

Ranking	Country	Region – NUTS 1	Total manufacturing (1000s)	% of which HRSTE	High and medium high technology (1000s)	% of which HRSTE	Medium low technology (1000s)	% of which HRSTE	Low technology (1000s)	% of which HRSTE
		EU-15 (2)	4 868	17.6	2 705	25.1	846	13.0	1 317	12.8
1	BE	Région Bruxelles-capitale	11	37.3	6	59.3	: u	: u	4	24.7 u
2	UK	London	82	34.5	33	43.5	: u	: u	43	35.9
3	FR	Île de France	177	33.9	106	40.0	15	17.8	56	32.4
4	ES	Comunidad de Madrid	98	32.9	60	47.5	13	23.3	25	21.8
5	ES	Noreste	135	32.1	55	38.6	49	31.4	31	25.6
6	UK	South East	150	30.8	99	37.6	15	19.3	36	24.7
7	DE	Berlin	43	29.6	27	34.9	: u	: u	11	24.4
8	FI	Manner-Suomi	121	28.9	61	38.3	19	21.9	40	23.7
9	DE	Hamburg	29	27.7	13	27.8	: u	: u	14	31.7
10	UK	Scotland	75	26.9	45	36.8	: u	: u	21	18.6
11	DE	Brandenburg	32	24.9	16	33.9	9	20.7	:	: u
12	IE	Ireland	57	24.7	34	35.0	7	16.7	16	17.3
13	ES	Noroeste	66	24.6	31	40.4	18	26.3	18	14.1
14	DE	Sachsen	75	24.3	34	27.4	15	20.9	26	23.3
15	BE	Vlaams Gewest	116	24.3	52	28.3	23	20.3	41	22.6
16	BE	Région Wallonne	40	24.0	21	38.2	12	21.0	7	13.0
17	DE	Baden-Württemberg	336	23.3	226	27.6	44	17.0	65	18.2
18	DE	Sachsen-Anhalt	34	22.5	19	30.2	9	19.6	: u	: u
19	DE	Thüringen	43	22.5	25	30.7	10	18.3	: u	: u
20	DE	Hessen	129	22.2	86	26.7	20	16.6	24	16.8
21	UK	North West (incl. Merseyside)	111	22.1	67	31.8	25	19.4	20	11.9
22	FR	Centre-Est	121	21.7	64	31.2	27	15.4	30	16.9
23	UK	Eastern	82	21.3	54	29.3	10	15.2	18	13.5
24	DE	Rheinland-Pfalz	82	21.2	47	24.0	14	17.2	20	19.2
25	ES	Este	220	21.1	113	35.9	33	14.8	74	14.6
26	ES	Canarias (ES)	8	20.8	: u	: u	: u	: u	4	19.1 u
27	DE	Bayern	298	20.8	198	26.9	31	12.5	68	15.4
28	UK	South West	62	20.4	39	25.2	10	21.5	13	12.5
29	DE	Schleswig-Holstein	33	20.3	21	25.2	: u	: u	8	14.7
30	FR	Sud-Ouest	73	20.0	49	36.1	8	12.3 u	16	9.8

(1) **Exceptions to the reference year 2002**
UK: 2000;
IS: 2001.
(2) EU-15: Eurostat estimates.

Source: Eurostat, EU LFS — spring data.

Ranked according to the number of employed people with tertiary education working in services, the leading region in the EU was again Région Bruxelles-capitale (BE) at 52.5%, this time followed by Spanish Noreste at 45.7% — Table 4.12. A further five Spanish regions were placed in the top 30 as well as six German and British, three Belgian and two French and Greek regions.

Table 4.12.

The proportion of S&T in services the top 30 regions in % 2002 (1)

Ranking	Country	Region — NUTS 1	Services (1000s)	% of which HRSTE	Knowledge intensive high tech services (1000s)	% of which HRSTE	Knowledge intensive financial services (1000s)	% of which HRSTE	Knowledge intensive market services (1000s)	% of which HRSTE	Other knowledge intensive services (1000s)	% of which HRSTE
		EU-15 (2)	29 383	30.5	2 027	38.6	1 609	32.8	4 254	37.3	13 115	47.9
1	BE	Région Bruxelles-capitale	138	52.5	11	61.7	11	67.0	28	54.4	47	67.0
2	ES	Noreste	414	45.7	17	65.0	22	57.6	53	51.5	158	70.3
3	ES	Comunidad de Madrid	682	44.0	94	68.8	60	59.7	136	51.9	188	62.8
4	FI	Manner-Suomi	606	43.3	51	52.1	26	60.3	82	44.4	264	51.0
5	UK	London	1 110	43.1	87	42.9	104	46.0	256	52.0	423	56.3
6	FR	Île de France	1 572	41.7	212	58.2	123	49.2	307	45.3	515	58.0
7	DE	Sachsen	407	40.2	18	45.4	19	47.9	52	39.9	174	59.2
8	BE	Région Wallonne	333	39.4	17	38.9	24	61.8	35	52.4	175	61.6
9	DE	Thüringen	219	39.2	: u	: u	9	45.4	20	45.4	96	57.0
10	BE	Vlaams Gewest	606	38.8	42	45.6	49	55.2	83	53.0	286	59.7
11	UK	Scotland	570	38.4	33	46.0	39	40.3	77	44.0	280	56.0
12	DE	Mecklenburg-Vorpommern	151	37.7	: u	: u	: u	: u	12	35.0	65	55.7
13	UK	Northern Ireland	157	37.6	: u	: u	: u	: u	12	37.0	79	61.5
14	DE	Berlin	388	37.6	28	41.9	14	32.8	58	38.1	163	50.1
15	IE	Ireland	338	36.7	32	49.5	28	49.4	50	45.9	149	54.6
16	DE	Brandenburg	238	36.6	9	26.4	11	45.3	26	40.0	101	56.7
17	NL	West-Nederland	859	36.4	72	48.8	52	37.1	159	40.3	356	47.7
18	ES	Noroeste	304	36.0	13	52.7	17	49.7	37	45.7	127	67.0
19	DK	Denmark	604	35.7	42	36.1	21	25.2	71	38.3	339	49.5
20	ES	Centro (E)	364	35.5	14	51.2	16	47.4	35	40.6	157	65.9
21	ES	Sur	599	35.2	24	56.0	31	52.0	63	40.7	265	66.5
22	FR	Centre-Est	611	35.1	45	46.2	31	43.2	76	41.8	289	50.9
23	UK	South East	921	35.1	105	43.9	52	30.1	151	38.7	393	51.1
24	DE	Sachsen-Anhalt	208	35.1	: u	: u	: u	: u	15	36.0	93	55.9
25	ES	Este	909	35.0	50	52.4	45	44.9	146	44.4	377	67.0
26	EL	Attiki	359	34.5	18	45.6	22	39.4	64	52.9	135	63.6
27	EL	Voreia Ellada	200	34.4	3	26.5 u	6	42.8	31	57.0	99	69.6
28	SE	Sweden	972	34.3 p	86	41.6 p	25	29.9 p	117	29.1 p	532	46.2 p
29	UK	Wales	245	33.3	: u	: u	: u	: u	31	43.5	149	50.1
30	UK	South West	486	32.1	31	34.5	25	28.9	76	39.4	229	49.2

(1) **Exceptions to the reference year 2002**
UK: 2000;
IS: 2001.
(2) EU-15: Eurostat estimates.

Source: Eurostat, EU LFS — spring data.

PART 3

European technological productivity and competitiveness

5.1. Introduction

Transforming technological knowledge into economic growth and welfare through its exploitation is a key tool to boost the competitiveness of a country in the modern economy. Being a complex phenomenon, evaluating how countries perform in developing and commercialising technology is not an easy task.

Among the indicators that may help to measure a country's performance in technological output, patents statistics are a widely used measure, as they represent an outcome of technologically oriented inventive activity. Although patents do not cover all kinds of innovation, they do account for a considerable part of it. There are some good reasons that have made patents one of the most widely used sources of data to construct indicators of inventive output such as the availability of detailed information for a relatively long time series or their close link to invention ([1]).

Nevertheless, using patent indicators does also have several shortcomings, and therefore patent indicators should be complemented with other S&T indicators so as to obtain a complete view of the innovation activities of the countries and regions. Among the drawbacks, an important one is that not all inventions are patented and not all patents have the same value. In fact, there is broad recognition that the value distribution of patents is very skewed: a few patents have large value, whereas many have very low value. However, as there are no generally recognised and easy applicable methods for measuring the value of patents, this chapter limits itself to simply counting the number of patents.

This chapter analyses the structure and evolution of patenting in the EU, Iceland, Liechtenstein, Norway, Candidate Countries, Japan and the United States and it is divided into three sections: a first section studies the international performances of EU-15, Japan and the United States, by looking at patent applications to the European Patent Office – EPO, patents granted by the United States Patent and Trademark Office – USPTO – as well as the so-called ***triadic patent families***; a second section focuses on the performance of European countries mainly at the EPO; finally, an insight into the patenting activities of the EEA regions at the EPO is given. The analysis covers the period from 1991 to 2001 for the EPO and USPTO data, whereas triadic patent family data covers the time-series from 1987 to 1998.

Patents statistics are very sensitive to the type of data collected and to the method used to count the patents. Therefore, data should be interpreted with caution, taking the following remarks into account:

The data presented in this chapter originate from three main sources:

- On the one hand, data on patent applications to the EPO were extracted from the EPO's database and have been processed by Eurostat.
- On the other hand, data on patents granted by the USPTO have been extracted from the USPTO's database and treated by the Fraunhofer ISI – FhG-ISI. Triadic patent family data originate from the OECD, who constructs such indicators combining data from the EPO, the USPTO and the Japanese Patent Office – JPO.
- In addition, some indicators from the ***Third European Report on Science and Technology Indicators*** – REIST-3 – prepared by the Directorate-General for Research are also presented in this chapter.

It should be noticed that EPO data refer to patent applications by year of filing, whereas USPTO data concern patents granted by year of publication. Although not all applications are granted, each application still represents technical effort by the inventor and therefore patent applications can be considered as an appropriate indicator of inventive potential. It takes on average just over four years for a patent to be granted at the EPO. In an effort to provide timely data therefore, Eurostat has chosen patent applications over patents granted. In the United States, up until recently, only information on granted patents was published and therefore data on applications is not yet presented in this chapter. In the USPTO, patents take from two to five years to be granted. With regard to the triadic patent families, they are counted according to the year of priority, i.e. year in which the patent was first applied for at any patent office and refer to applications to the EPO and the JPO and grants by the USPTO.

When interpreting the data at the international level, the reader should bear in mind that due to a 'home' advantage, European countries may be dominant in the European patent system, whereas the United States may be dominant in the US patent system. On the other hand, figures may also be influenced by the countries' industrial structures, as different industries have a different propensity to patent.

Some of these weaknesses are reduced or suppressed in the triadic patent family indicators, as they only take into account patents that have been applied for at the EPO and the JPO and granted by the USPTO. Besides improving international comparability of patent based indicators, triadic patent family data also balances the differences in the value of the patents associated with traditional indicators. This is because patenting in the three offices is very costly due to, not only administrative fees, but also translation costs. In this context, the patentee will only proceed to do such applications if he/she deems it worthwhile, i.e. if the expectation for having the patent granted and the expected return from protection through sales or licenses in the designated countries are high enough.

Due to methodological differences in the manner of processing the data, no cross sectional comparisons are advisable between the EPO, the USPTO and triadic patent family data. For further explanations on the methodology used, please refer to the methodological notes starting on page 150 or to Eurostat's reference database *NewCronos*.

(1) For further details on the advantages and drawbacks of using patents statistics as an S&T indicator, please refer to the methodological notes in Eurostat's reference database ***NewCronos*** or to ***Statistics on Science and Technology — 1991-2001***, Detailed tables collection, Theme 9 — Science and Technology, Eurostat, 2003.

5.2. International comparison: EU-15, Japan and the United States

The performance of the EU, Japan and the United States in technological output is analysed in this section by looking at their patenting activities at the EPO and the USPTO, as well as their attainment in terms of triadic patent families.

Patent applications to the EPO

Whilst the EU led in absolute terms, Japan retained the highest rate relative to population

As globalisation shows that protecting inventions is becoming increasingly important, patent applications to the EPO from the EU, Japan and the United States continue on an upward trend – Figure 5.2. Although patent applications to the EPO were already growing steadily during the first part of the nineties, annual average growth rates were especially high from 1996 onwards for all the EU, Japan and the United States. For the 1996-2001 period, it was Japan that recorded the highest annual average growth rate (11.9% per annum), followed by the EU (11.0%) and the United States (10.9%).

In 2001, the EPO received 60 890 patent applications from EU Member States, 168.3% of its value in 1996 and more than double the applications made in 1991. Patent applications to the EPO from Japan and the United States in 2001 amounted to 22 226 and 47 202, respectively. These represented 175.8% and 167.8% of their corresponding values in 1996. Europe still has the highest share of patents at the European Patent Office, but according to the *Third European Report on Science and Technology Indicators* – REIST-3, the United States has increased its presence over the past decade in terms of percentage of patent applications filed at the EPO. The gap between the EU and the United States was 16.6% in 1992, compared with only 9.8% in 2001. In 2001, the EU was responsible for 42.2% of patents applied for at the EPO, the United States for 32.4% and Japan for 14.6%. Over the period 1992 to 2001, the United States managed to increase its share by 4.2%, while the EU's share fell by 2.6% and Japan's by 4.9% – REIST-3, p.329.

Figure 5.3. shows the evolution of patent applications to the EPO as a proportion of population for the 1991-2001 period. When taking population into account, the differences across the three blocks become smaller and the positions invert. In 2001, the highest ratio was registered by Japan – 175 patent applications per million inhabitants, followed by the United States (170) and the EU (161).

When set against its business R&D effort, the EU's performance in patenting has been relatively healthy since the late 1990's. As shown in Figure 5.1., during the 1997-2001 period, the EU recorded a rate of European patents per unit of business R&D expenditure equal to 0.60, followed by Japan with 0.31 and the United States with 0.26 all countries increasing their rates with regard to the 1992-96 period. In spite of a much more modest increase in R&D spending by business compared with its counterparts, Europe has managed to generate a significant growth in patenting at the EPO. This may suggest that the EU patenting activity has been boosted by more than just an expansion of business research spending.

Figure 5.1. **European patents per unit of business R&D expenditure (1) EU-15, Japan and the United States 1992 to 1996 and 1997 to 2001**

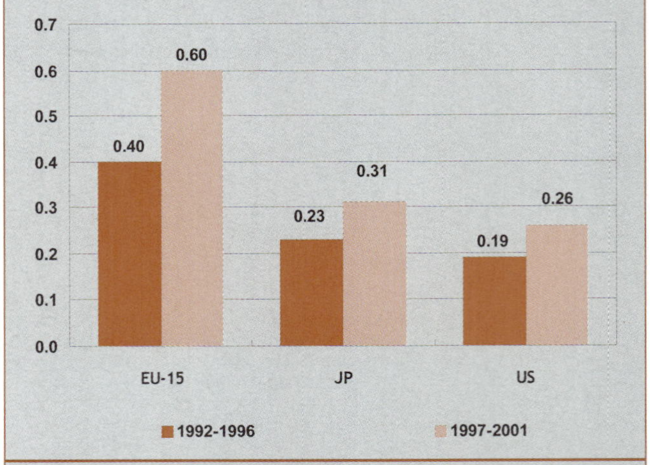

(1) Business R&D — BERD — measured in million purchasing power standards — PPS — at 1995 prices. Calculated using a two-year time lag between year of R&D expenditure and year of patenting.

Source: *Third European Report on Science and Technology Indicators*, Directorate-General for Research, 2003, p. 352.
Data: EPO — data processed by OST; OECD, Eurostat.

Figure 5.2.

Trends of patent applications to the EPO
EU-15, Japan and the United States
1991 to 2001 (1)

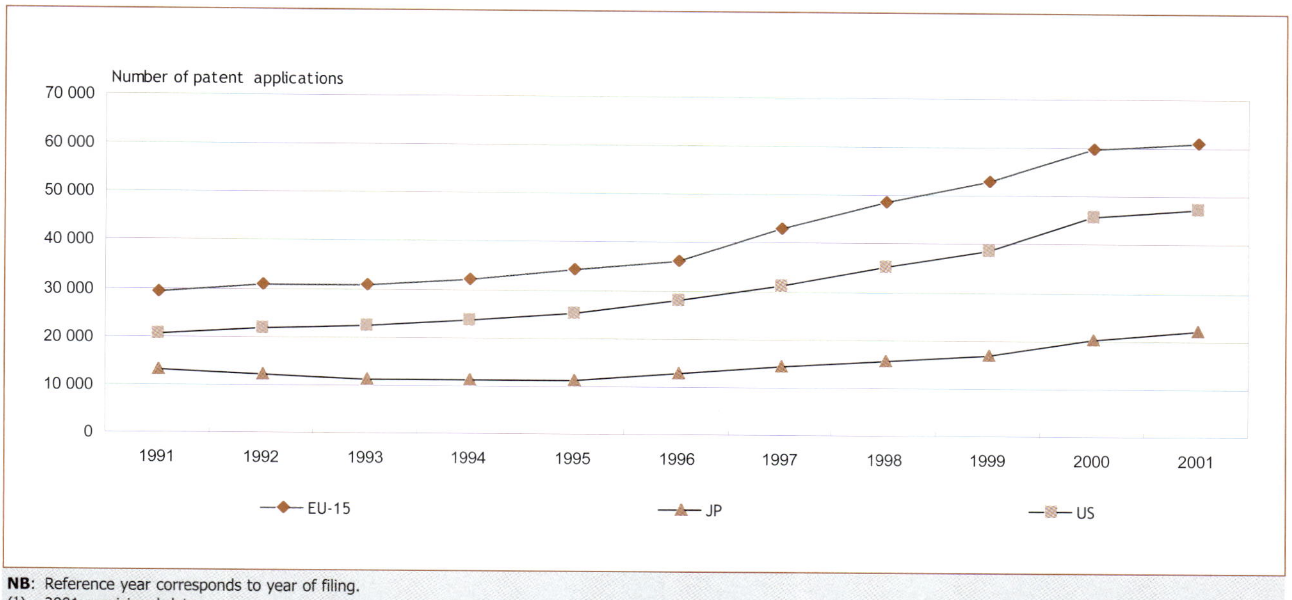

NB: Reference year corresponds to year of filing.
(1) 2001 provisional data.

Sources: Eurostat, EPO.

Figure 5.3.

Trends of patent applications to the EPO per million inhabitants
EU-15, Japan and the United States
1991 to 2001 (1)

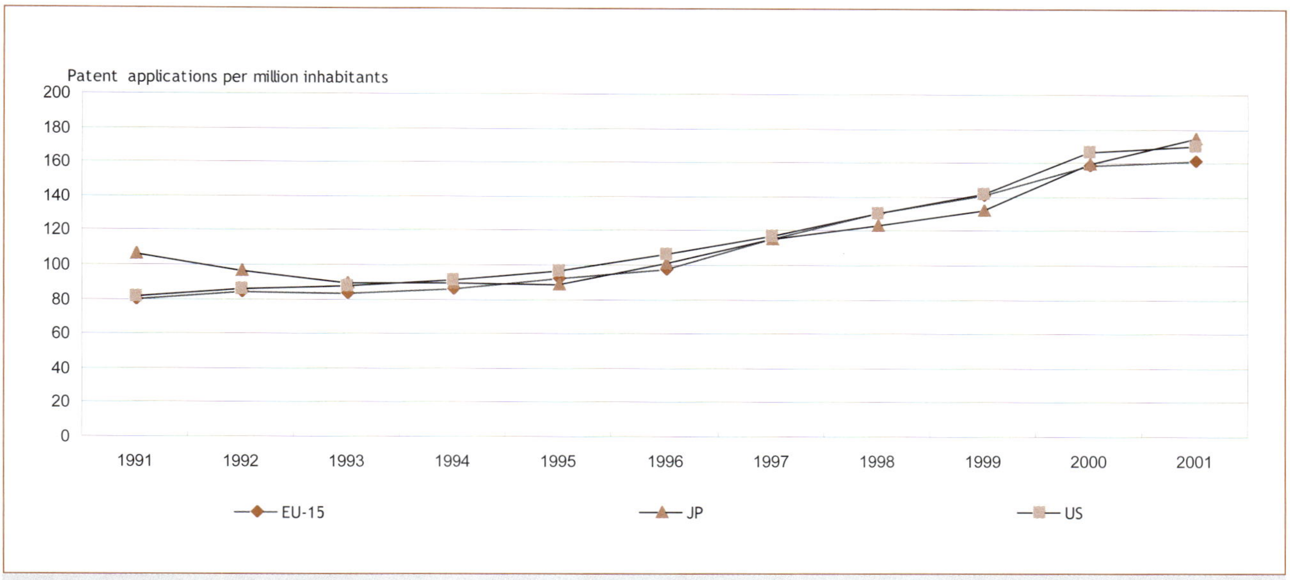

NB: Reference year corresponds to year of filing.
(1) 2001 provisional data;
EU-15 — 1999, 2000, 2001 population data: Eurostat estimates;
JP and US — 2001 population data: Eurostat estimates.

Sources: Eurostat, EPO.

Patents are classified according to the International Patent Classification, which is commonly referred to as IPC. According to the IPC classification, an invention is assigned to an IPC-class by its function or intrinsic nature, or by its field of application. The IPC is therefore a combined function-application classification system in which the function takes precedence.

Table 5.1. shows the distribution of patent applications to the EPO from the EU, Japan and the United States by IPC section. EU patent applications to the EPO in 2001 specialised in the 'Performing operations; transporting' section (19.4% of total applications). Whilst Japan specialised in 'Electricity' (26.4%), the United States applied for most patents on the field of 'Physics' (25.5%).

The second largest section in the EU and the United States was 'Electricity', as it accounted for 18.8% and 19.6% of their respective totals. 'Physics' was the second largest section for Japan (24.0%).

An increasing proportion of patent applications to the EPO refers to applications in the high technology fields – see definition of high tech patents in methodological notes starting on page 150. This increasing trend is clear for both the EU and the United States in Figure 5.4., as high tech patent applications to the EPO are plotted as a percentage of total applications. Whilst they represented 9.4% of total applications from the EU in 1991, they amounted to 19.6% in 2001. United States high tech applications increased from 18.5% in 1991 to 33.6% in 2001. The trend for Japan appears fairly stable ranging from 24.5% in 1991 to 25.7% in 2001.

The increasing proportion of high tech patents is explained by the fact that applications in this fields are growing faster than other type of applications. As shown in Table 5.2., whilst patent applications to the EPO from the EU in the 1996-2001 period grew at an annual average growth rate of 11.0%, applications in the high tech fields grew at a rate of 22.3% per annum. The growth rate for high tech applications from the United States (20.4%) also almost doubled that of patents overall (10.9%). Although the difference is not as striking, Japanese high tech patent applications (15.4%) also grew at a rate well above the total (11.9%).

In 2001, high tech patent applications to the EPO from the EU amounted to 11 928. The United States and Japan applied for 15 839 and 5 707 high tech patents, respectively. As a proportion of population, with 57 high tech patent applications per million inhabitants, the United States was ahead, followed by Japan (45) and the EU (32). Although the EU seems to be lagging behind its competitors in the high tech fields, it is also catching up, as shown by the annual average growth rates.

Table 5.3. shows the distribution of high tech patent applications to the EPO by high tech group. High tech patents may be grouped in the following technology groups (2):

- Aviation – AVI,
- Computer and automated business equipment – CAB,
- Communication technology – CTE,
- Lasers – LSR,
- Micro-organism and genetic engineering – MGE,
- Semi-conductors – SMC.

Most high tech patent applications to the EPO from EU-15 (47.1% of total applications) and Japan (38.1%) in 2001 were done in the field of 'Communication Technology', which includes electrical communication systems such as telephones or television. This is also the second largest group for the United States (30.8%). 'Computer and automated business equipment' was the most important high tech group for the United States (41.5%), whereas it was the second one for both the EU (28.5%) and Japan (35.0%).

(2) See composition of each group in methodological notes starting on page 150.

Table 5.1.

Distribution of patent applications to the EPO, by IPC section in %
EU-15, Japan and the United States
2001 (1)

IPC section		EU-15	JP	US
A	Human necessities	15.0	9.4	18.1
B	Performing operations; transporting	19.4	14.8	11.3
C	Chemistry; metallurgy	14.3	15.7	18.2
D	Textiles; paper	1.9	1.2	1.0
E	Fixed constructions	4.2	0.7	1.6
F	Mechanical engineering; lighting; heating; weapons; blasting	9.8	7.7	4.6
G	Physics	16.7	24.0	25.5
H	Electricity	18.8	26.4	19.6
Total number		60 890	22 226	47 202

NB: Reference year corresponds to year of filing.
(1) 2001 provisional data.

Sources: Eurostat, EPO.

Figure 5.4.

Trends of high tech patent applications to the EPO as a % of total applications
EU-15, Japan and the United States
1991 to 2001 (1)

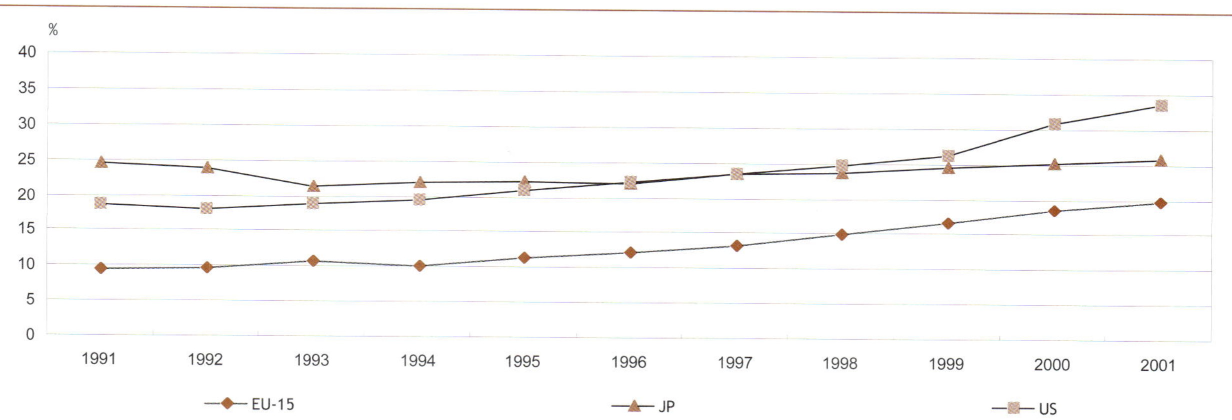

NB: Reference year corresponds to year of filing.
(1) 2001 provisional data.

Sources: Eurostat, EPO.

Table 5.2.

High tech patent applications to the EPO
EU-15, Japan and the United States
1991 to 2001 (1)

	EU-15	JP	US
High tech patent applications in 2001			
Total number	11 928	5 707	15 839
Per million inhabitants	32	45	57
As a % of total	19.6	25.7	33.6
Annual average growth rates in %			
High tech patents 1991-96	9.7	-2.9	10.5
High tech patents 1996-2001	22.3	15.4	20.4
All patents 1991-96	4.4	-0.8	6.6
All patents 1996-2001	11.0	11.9	10.9

NB: Reference year corresponds to year of filing.
(1) 2001 provisional data.
EU-15, JP and US — 2001 population data: Eurostat estimates.

Sources: Eurostat, EPO.

Table 5.3. **Distribution of high tech patent applications to the EPO, by high tech group in %**
EU-15, Japan and the United States
2001 (1)

High tech group	EU-15	JP	US
Aviation	1.2	0.3	0.8
Computer and automated business equipment	28.5	35.0	41.5
Communication technology	47.1	38.1	30.8
Lasers	1.4	2.3	1.5
Micro-organism and genetic engineering	13.0	8.4	17.0
Semi-conductors	8.9	15.9	8.4
Total number	11 928	5 707	15 839

NB: Reference year corresponds to year of filing.
(1) 2001 provisional data.

Sources: Eurostat, EPO.

Patents granted by the USPTO

Great differences between the number of domestic patents and foreign ones

As regards industrial property protection in the United States, the number of patents granted by the USPTO is also on an upward trend – Figure 5.6. Among the triad, the EU recorded the highest annual average growth rate of patents granted by the USPTO for the 1996-2001 period (10.8%). However, this can just be seen as a certain consolidation, as comparing the share of the EU in the total number of patents granted by the USPTO, no increase can be noticed between 1991 and 2001. During the 1996-2001 period, the number of patents granted to both US and Japan grew at an annual average growth rate of 7.7% per annum. The differences between the number of domestic patents and foreign ones is indeed somewhat striking. Whilst the USPTO granted 89 636 patents to US inventors in 2001, only 30 285 were awarded to inventors from the EU and 33 733 to inventors from Japan.

As shown in Figure 5.7., when taking population into account, differences still remain large at the USPTO. In 2001, the United States retained a ratio of 322 patents granted per million inhabitants. With 265 patents granted per million inhabitants, the gap with Japan when taking population into account reduces. The ratio of patents granted to the EU per million inhabitants in 2001 was 80, one fourth of the United States' rate.

It may be argued whether the position of EU Member States in the United States and Japan is comparable to that of the United States or Japan in the EU. This is directly linked to the complexity of the European patenting scenario, where the European patenting system (3) coexists with those of the Member States. This has implications on the cost of patenting in Europe, which has been proved to be three to five times more expensive than in the United States or Japan. The Commission estimated that whilst the overall cost of an European patent including translation costs and other fees is around EUR 49 900, Japanese and US patents cost on average EUR 16 450 and EUR 10 330, respectively (4).

(3) Please note that an European patent does not necessarily imply protection in the entire EU territory, but only at the designated states. This is not the case for the United States or Japanese patents, where one patent always covers the whole country.

(4) See ***Proposal for a Council Regulation of the Community patent,*** Commission of the European Communities, Brussels 1.8.2000, COM(2000)412 final.

Figure 5.1. set European patents against its business R&D effort, showing a relatively healthy performance of the EU since the late 1990's.

Figure 5.5. looks at the corresponding figures for the US patents. During the 1997-2001 period, the United States recorded the highest rate of US patents per unit of business R&D expenditure (0.58), followed by Japan with 0.55 and the EU with 0.33. As for the European patents, all blocks increased their rates with regard to the 1992-96 period.

Figure 5.5. **US patents per unit of business R&D expenditure (1) 1992 to 1996 and 1997 to 2001**

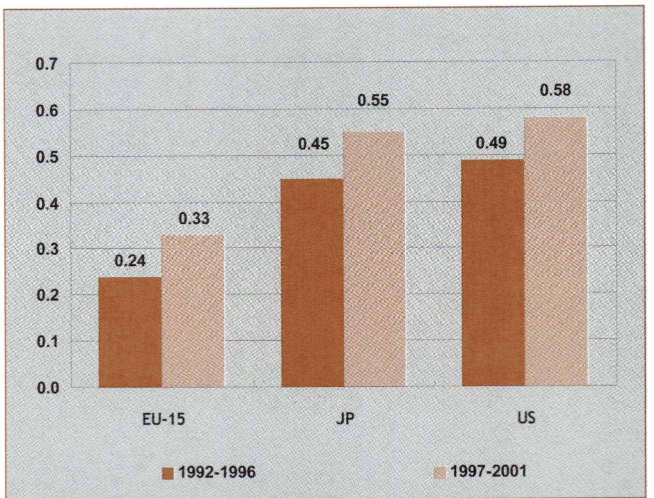

(1) Business R&D — BERD — measured in million purchasing power standards — PPS — at 1995 prices. Calculated using a two-year time lag between year of R&D expenditure and year of patenting.

Source: *Third European Report on Science and Technology Indicators,* Directorate-General for Research, 2003, p. 353.
Data: EPO — data processed by OST; OECD, Eurostat.

Figure 5.6.

Trends of patents granted by the USPTO
EU-15, Japan and the United States
1991 to 2001

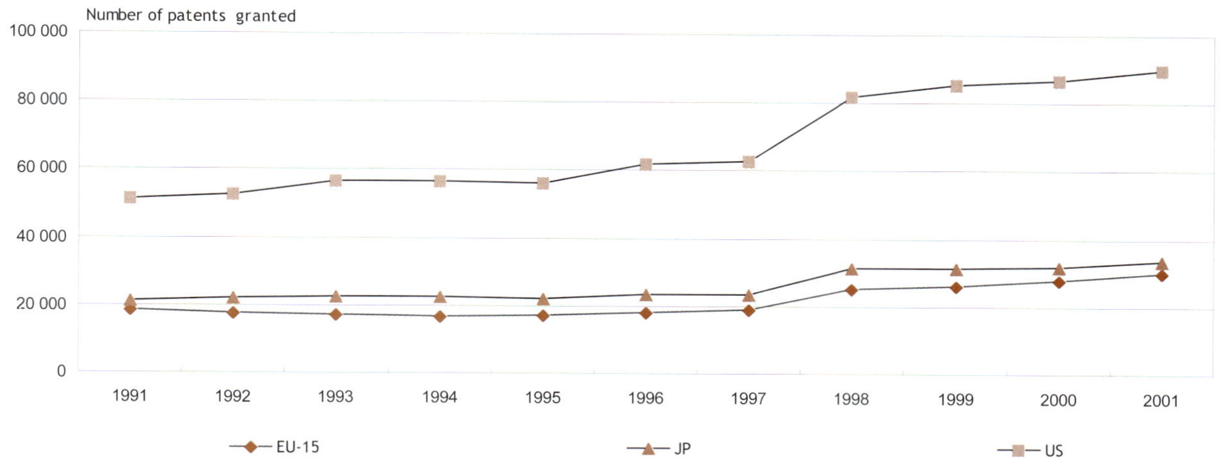

NB: Reference year corresponds to year of publication.

Sources: Eurostat, USPTO.

Figure 5.7.

Trends of patents granted by the USPTO per million inhabitants
EU-15, Japan and the United States
1991 to 2001 (1)

NB: Reference year corresponds to year of publication.
(1) EU-15 — 1999, 2000, 2001 population data: Eurostat estimates;
JP and US — 2001 population data: Eurostat estimates.

Sources: Eurostat, USPTO.

Triadic patent families

In absolute terms, the United States leads closely followed by the EU; as a proportion of population Japan is ahead

In order to overcome comparability problems associated to data derived from patents filed at a single patent office, the OECD is using the concept of patent family. A patent family is defined as a set of patents taken in various countries for protecting a single invention. In other words, a patent is a member of the patent families if, and only if, it is filed at the European Patent Office – EPO, the Japanese Patent Office – JPO – and is granted by the United States Patent and Trademark Office – USPTO.

As patent families are counted according to the priority year, i.e. year in which the patent was first applied for at any patent office, the latest year for which triadic patent family data are available is 1998. This is because the time lag between the priority date and the availability of information on patent applications to the EPO and JPO could be up to 4 years. As a triadic patent family is only counted after the USPTO has granted it, the duration of this granting procedure and its publication also needs to be taken into account. In total, therefore, information on USPTO grants could be available up to 6 to 10 years after the priority date. Hence, at present the OECD has almost complete patent families data up to 1996 only. In this context, data for 1997 and 1998 are OECD Secretariat estimates based on projections of the number of USPTO patent grants, evaluated using the available data for these years and the time lags between priority and grant over the period 1992-96.

The evolution of the number of triadic patent families in the EU, Japan and the United States is revealed in Figure 5.8. In 1998, the patentees from the United States registered the highest number of triadic patent families (14 255), closely followed by the EU (13 187) and Japan (10 033). It may be seen that compared to the figures for each individual patent office, when only patent families are taken into account, differences across the three blocks are somewhat reduced.

Although the upwards trend in the case of patent families is not as clear and steady as it is for individual patent offices, all the EU, Japan and the United States recorded positive annual average growth rates for the 1987-98 period. A steady upward trend is especially visible from 1993 onwards. For the 1993-98 period, the EU recorded the highest annual average growth rate of triadic patent families (6.2% per annum), followed by the United States (5.2%) and Japan (3.9%).

The distribution of triadic patent families in the OECD's total is shown in Figures 5.9. and 5.10. Although the EU accounted for the largest proportion of triadic patent families in 1987 and 1988, the United States has retained a larger proportion since, with the exception of the year 1996, when both the EU and the United States accounted for 34% of the total. The proportion of patent families by Japanese inventors increased in the late eighties, but is on a downward trend since 1990. In 1998 the United States accounted for the largest percentage of patent families in the OECD (35.7%), followed by the EU (33.5%) and Japan (25.4%).

When taking population into consideration, Japan leads ahead of the United States and the EU. As shown in Figure 5.11., during the 1987-98 period, Japanese triadic patent families per million inhabitants were well above the corresponding ratios retained by the United States and the EU. In 1998, Japan registered 81 patent families per million inhabitants, followed by the United States with 54 and the EU with 36.

Figure 5.8. **Trends of triadic patent families EU-15, Japan and the United States 1987 to 1998 (1)**

NB: Reference year corresponds to year of priority.
(1) **OECD Secretariat estimates or projections based on national sources**
EU-15, JP and US: 1997 and 1998;
EU-15 and JP: 1996.

Source: OECD MSTI 2003/1.

Figure 5.9. **Distribution of triadic patent families in OECD total EU-15, Japan and the United States 1987 to 1998 (1)**

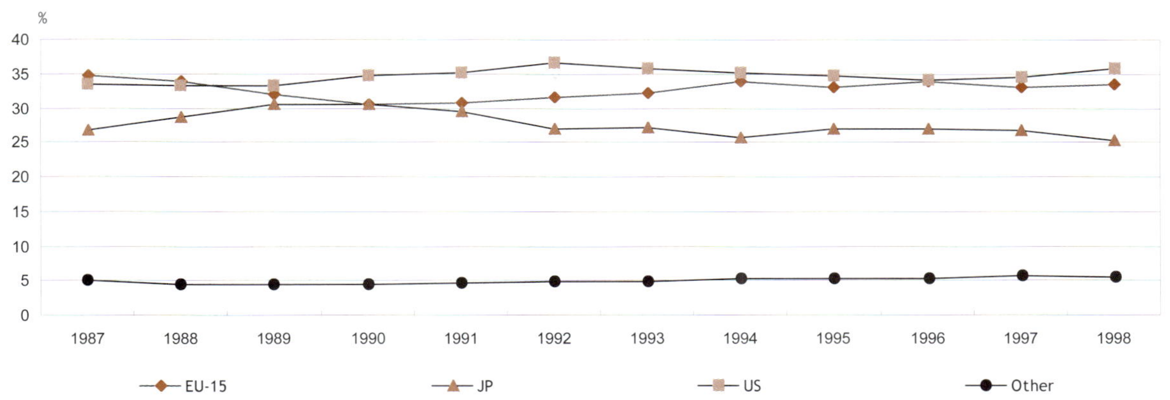

NB: Reference year corresponds to year of priority.
(1) **OECD Secretariat estimates or projections based on national sources**
EU-15, JP and US: 1997 and 1998;
EU-15 and JP: 1996.

Source: OECD MSTI 2003/1.

Figure 5.10. **Distribution of triadic patent families in OECD total EU-15, Japan and the United States 1998 (1)**

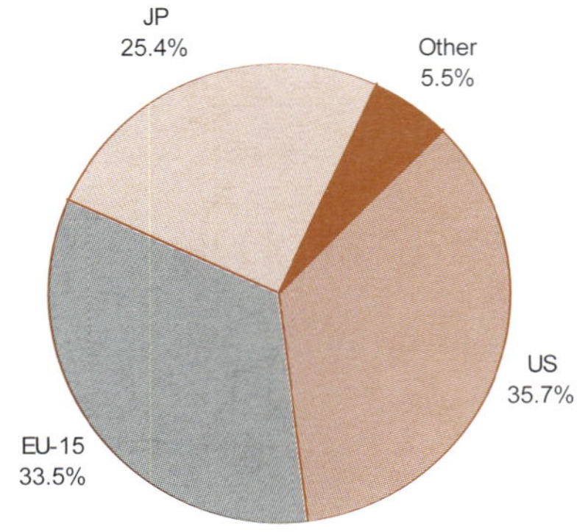

NB: Reference year corresponds to year of priority.
(1) **OECD Secretariat estimates or projections based on national sources**

Source: OECD MSTI 2003/1.

Figure 5.11. **Trends of triadic patent families per million inhabitants EU-15, Japan and the United States 1987 to 1998 (1)**

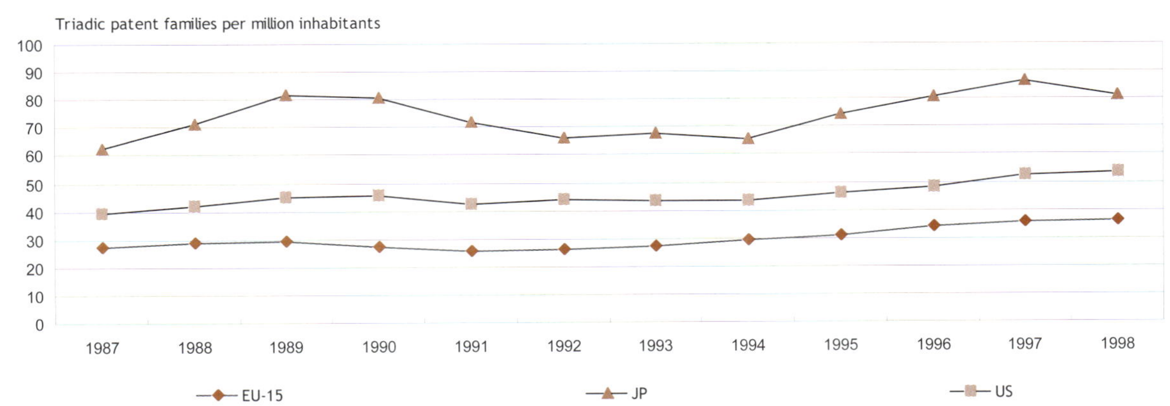

NB: Reference year corresponds to year of priority.
(1) EU-15 — 1999, 2000, 2001 population data: Eurostat estimates; JP and US — 2001 population data: Eurostat estimates.
OECD Secretariat estimates or projections based on national sources
EU-15, JP and US: 1997 and 1998; EU-15 and JP: 1996.

Sources: Eurostat; OECD MSTI 2003/1.

Table 5.4.

Patent applications to the EPO
EU-15, Candidate Countries, Iceland, Liechtenstein and Norway
2001 (1)

	Total number	Per million inhabitants	Distribution by IPC section in % (6)							
			A	B	C	D	E	F	G	H
EU-15 (2)	**60 890**	**161**	**15.0**	**19.4**	**14.3**	**1.9**	**4.2**	**9.8**	**16.7**	**18.8**
EUR-12 (2)	**48 516**	**160**	**14.2**	**20.8**	**14.4**	**1.9**	**4.2**	**10.4**	**15.5**	**18.5**
BE	1 558	152	14.9	18.5	28.5	3.1	3.8	6.1	13.6	11.4
DK	1 129	211	24.1	13.0	18.4	1.1	4.6	8.7	16.1	14.1
DE	25 489	310	11.7	22.3	14.8	1.7	4.1	12.6	15.4	17.4
EL (2)	82	8	22.8	16.1	11.7	-	4.3	11.0	15.3	18.7
ES	967	24	23.3	22.4	14.7	2.1	5.9	7.9	11.6	12.2
FR	8 580	145	18.3	19.0	13.9	1.1	3.7	8.1	16.9	18.9
IE (4)	327	86	21.5	13.4	8.7	0.2	3.9	3.7	25.9	22.6
IT (3)	4 318	75	21.1	27.1	11.3	3.4	5.5	11.2	9.7	10.8
LU (3)	93	211	2.8	28.4	21.5	-	9.1	21.4	8.7	8.0
NL	3 881	243	13.5	12.3	14.6	0.8	3.1	4.7	22.4	28.8
AT	1 414	174	13.6	22.2	14.6	2.3	9.1	11.4	11.6	15.1
PT (3)	56	5	24.1	20.2	22.6	1.2	8.4	11.9	5.3	6.2
FI	1 750	338	7.8	13.9	6.8	7.6	2.7	5.0	15.5	40.6
SE	3 256	367	15.8	17.6	7.6	2.7	2.9	8.9	17.3	27.3
UK (3)	7 989	133	18.1	12.6	15.7	1.2	4.7	6.3	23.4	18.1
ACC (5)	**568**	**8**	**24.3**	**13.0**	**15.7**	**2.0**	**4.1**	**9.0**	**16.3**	**15.6**
CZ	110	11	17.6	24.3	15.3	8.2	3.6	11.4	12.4	7.2
EE (4)	15	11	23.4	6.6	10.0	-	-	13.3	40.1	6.7
CY	11	14	45.4	9.1	18.2	-	-	-	9.1	18.2
LV	18	8	37.3	9.2	38.1	-	4.2	5.6	5.6	-
LT	9	2	26.7	-	51.1	-	-	-	22.2	-
HU (2)	190	19	24.3	8.3	13.0	-	3.2	4.9	24.2	22.1
MT	4	10	8.3	-	-	-	-	50.1	41.6	-
PL	97	3	23.7	17.0	12.3	1.9	7.2	14.2	12.6	11.1
SI	81	41	29.8	8.1	17.4	0.6	6.8	6.2	8.5	22.6
SK	33	6	23.0	14.6	19.9	-	-	16.7	6.1	19.7
BG	17	2	37.3	17.7	5.8	-	-	17.6	21.6	-
RO	17	1	8.8	23.5	17.7	-	7.4	11.8	1.5	29.4
TR (2)	72	1	30.6	4.2	7.6	10.4	3.5	17.4	11.1	15.3
EEA (2)	**62 259**	**163**	**15.1**	**19.4**	**14.2**	**1.8**	**4.3**	**9.8**	**16.7**	**18.7**
IS	33	117	53.8	4.5	17.7	-	-	-	13.5	10.5
LI (3)	36	1 080	29.4	27.5	14.9	-	7.0	11.2	3.6	6.3
NO	1 300	289	20.8	16.2	11.2	0.4	11.7	10.4	17.8	11.4

NB: Reference year corresponds to year of filing.
(1) 2001 provisional data.
(2) EU-15, EUR-12,EL, HU, TR and EEA — 2001 population data: Eurostat estimates.
(3) IT, LU, PT, UK and LI — 2001 population data: estimated values.
(4) IE and EE — 2001 population data: provisional value.
(5) Acceding Countries — ACC — includes: CZ, EE, CY, LV, LT, HU, MT, PL, SI and SK. 2001 population data: Eurostat estimates.
(6) A Human necessities;
B Performing operations; transporting;
C Chemistry; metallurgy;
D Textiles; paper;
E Fixed constructions;
F Mechanical engineering; lighting; heating; weapons; blasting;
G Physics;
H Electricity.

Sources: Eurostat, EPO.

5.3. Performance at the national level in Europe

Total patent applications to the EPO

Germany leads at the EPO in absolute terms, whereas Sweden and Finland are ahead as a proportion of population

At the EU Member State level, Germany leads in absolute terms, with 25 489 patent applications in 2001 – Table 5.4. Following Germany were France (8 580) and the United Kingdom (7 989). Nevertheless, when taking population into account, with 367 and 338 patent applications per million inhabitants respectively, Sweden and Finland are ahead of Germany which retained a rate of 310 – Figure 5.12.

Although patent applications to the EPO from Acceding and Candidate Countries are still below the European Union average, Hungary (190),the Czech Republic (110 patent applications), Poland (97), Slovenia (81) and Turkey (72) applied for a higher number than the EU Member State that applied for the least amount of patents at the EPO (Portugal, 56). As a proportion of population, Slovenia was the Acceding Country that retained the highest rate (41 patent applications per million inhabitants), Followed by Hungary (19) and Cyprus (14). Eight Acceding Countries registered rates per million inhabitants above that of Portugal and six above that of Greece – Table 5.4.

Although on average the EU patent applications to the EPO in 2001 specialised in 'Performing operations; transporting – Section B' with 19.4% of the total applications, the distribution across IPC sections varies at the Member State level: 4 countries specialised in 'Human necessities – Section A', 5 in 'Performing operations; transporting – Section B', 1 in 'Chemistry; metallurgy – Section C', 2 in 'Physics – Section G' and 3 in 'Electricity – Section H' – Table 5.4.

Although the distribution across IPC sections varied among Acceding Countries, overall, they applied for most patents in the 'Human necessities – Section A' field (24.3% of total applications), which is the 4th largest section in the EU total. 'Performing operations; transporting – Section B', the largest section for the EU, was only the fifth in the Acceding Countries out of a total of 8 sections.

Figure 5.12. **Patent applications to the EPO per million inhabitants EU-15, Iceland and Norway 1991, 1996 and 2001 (1)**

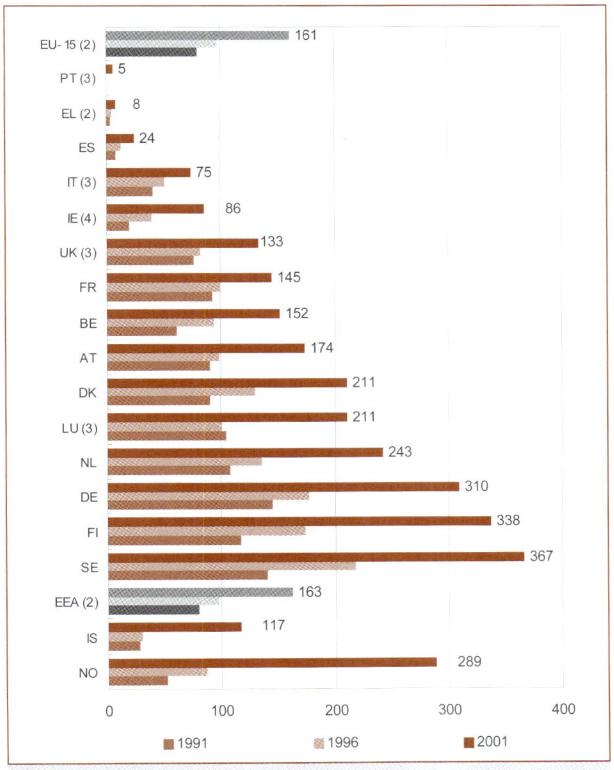

NB: Reference year corresponds to year of filing.

(1) 2001 provisional data.
(2) EU-15, EL and EEA — 2001 population data: Eurostat estimates.
(3) IT, LU, PT and UK — 2001 population data: estimated values.
(4) IE — 2001 population data: provisional value.

Sources: Eurostat, EPO.

Figure 5.13. **Patent applications to the EPO per million inhabitants Candidate Countries 1991, 1996 and 2001 (1)**

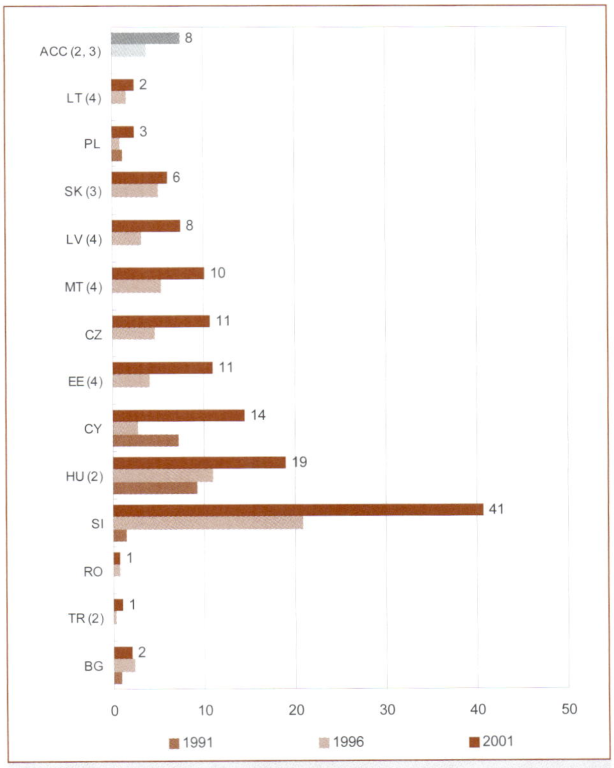

NB: Reference year corresponds to year of filing.

(1) 2001 provisional data.
(2) ACC, HU and TR — 2001 population data: Eurostat estimates.
(3) ACC and SK — 1991: not available.
(4) EE, MT, LV and LT — 1991: value equal to real zero.

Sources: Eurostat, EPO.

As regards the distribution of patents among EU Member States, a strong skewness towards the large European Economies may be observed both in terms of the total number of patent applications to the EPO and grants by the USPTO – Figures 5.14. and 5.15.

At the EPO, Germany accounted for the largest amount of patent applications (41.9%), followed by France (14.1%) and the United Kingdom (13.1%). Together they represented over two thirds of the EU total.

The distribution of patents granted by the USPTO is similar: Germany leads with 40.0% of the patents granted, but the United Kingdom (15.3%) is ahead of France (14.9%). Germany, the United Kingdom and France together they also represent over two thirds of the total number of patents granted by the USPTO to EU inventors.

High tech patent applications to the EPO

High tech patents account for an increasing proportion of applications

Although patent applications to the EPO are overall growing for all the Member States of the EU, this is especially true for patents in the high technology fields. As shown in Table 5.5., patent applications to the EPO in the high technology fields during the 1996-2001 period grew at an annual average growth rate that doubled that of patents overall: whilst all patents grew at 11.0% per annum, high tech patents grew at 22.3%. At the Member State level, from 1996 to 2001, high tech patent applications grew faster than patents overall for all countries except Italy.

On average, in 2001 high tech patents accounted for 19.6% of the total number of applications from EU inventors. The EU Member State for which high tech patents accounted for the highest proportion of the total was Finland (40.3%), followed by Ireland (35.9%) and the Netherlands (28.3%).

Figure 5.14. **Distribution of patent applications to the EPO EU-15 by Member State 2001**

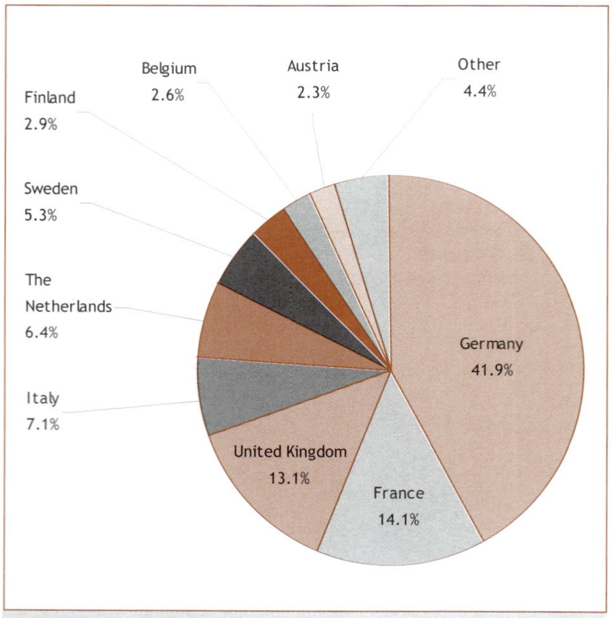

NB: Reference year corresponds to year of filing.
(1) 2001 provisional data.

Sources: Eurostat, EPO.

Figure 5.15. **Distribution of patents granted by the USPTO EU-15 by Member State 2001**

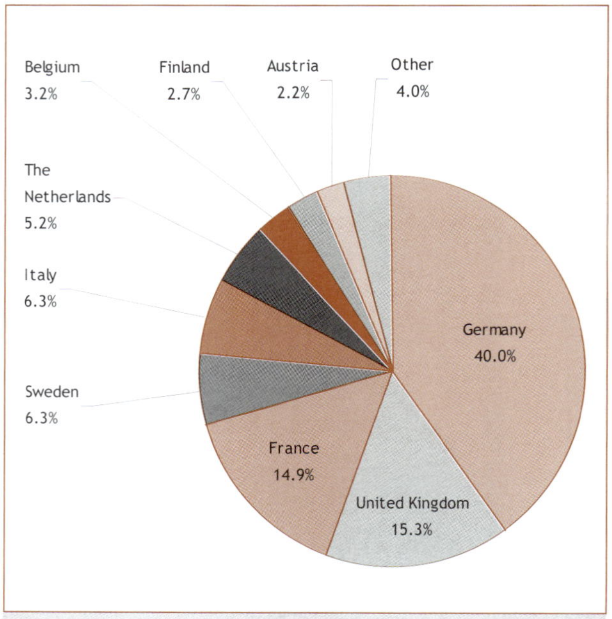

NB: Reference year corresponds to year of publication.

Sources: Eurostat, USPTO.

Table 5.5.

High tech patent applications to the EPO
EU-15, Candidate Countries, Iceland, Liechtenstein and Norway
1991 to 2001 (1)

	High tech patent applications in 2001			Annual average growth rates in %			
	Total number	Per million inhabitants	As a % of total	High tech patents 1991-96	High tech patents 1996-2001	All patents 1991-96	All patents 1996-2001
EU-15 (2)	**11 928**	**32**	**19.6**	**9.7**	**22.3**	**4.4**	**11.0**
EUR-12 (2)	**8 673**	**29**	**17.9**	**9.1**	**22.2**	**4.5**	**11.0**
BE	240	23	15.4	12.9	14.0	9.3	10.3
DK	225	42	19.9	9.5	27.1	8.1	10.6
DE	4 017	49	15.8	8.8	24.6	4.7	11.9
EL (2)	22	2	27.0	3.1	50.4	7.2	11.4
ES	143	4	14.8	10.3	31.5	9.0	13.6
FR	1 791	30	20.9	3.0	20.1	1.9	8.2
IE (4)	117	31	35.9	20.5	36.0	15.7	18.1
IT (3)	374	6	8.7	12.6	6.8	5.0	8.3
LU (3)	5	11	5.1	-	48.6	0.7	17.5
NL	1 100	69	28.3	11.4	25.2	5.4	13.0
AT	152	19	10.8	7.5	20.1	2.3	12.3
PT (3)	7	1	12.3	2.2	64.8	11.7	30.0
FI	705	136	40.3	31.3	23.1	8.8	14.5
SE	896	101	27.5	29.5	23.0	9.8	11.1
UK (3)	2 134	36	26.7	7.6	21.7	2.0	10.6
ACC (2)	**89**	**1**	**15.6**	**:**	**37.0**	**:**	**14.8**
CZ	7	1	6.2	-	9.4	116.9	18.0
EE (4)	2	1	13.4	-	31.4	-	20.0
CY	2	3	18.2	-	-	-16.7	40.6
LV	1	0	5.6	-	-	-	17.5
LT	3	1	28.8	-	-	-	8.4
HU (2)	43	4	22.9	7.3	40.1	3.1	11.2
MT	-	-	-	-	-	-	14.8
PL	8	0	7.8	-11.5	72.5	-6.2	25.0
SI	17	9	21.4	-	63.3	69.1	14.3
SK	6	1	17.9	:	18.9	:	4.1
BG	3	0	17.6	-3.0	14.8	22.1	-2.2
RO	3	0	17.7	0.0	24.6	41.4	0.0
TR (2)	12	0	16.7	-	59.3	20.9	31.9
EEA (2)	**12 160**	**32**	**19.5**	**9.8**	**22.6**	**4.5**	**11.2**
IS	9	31	26.5	-17.9	77.4	2.8	32.7
LI (3)	-	-	-	-	-	-2.4	3.4
NO	223	50	17.2	23.2	58.5	11.6	27.9

NB: Reference year corresponds to year of filing.

(1) 2001 provisional data.
(2) EU-15, EUR-12, EL, ACC, HU, TR and EEA — 2001 population data: Eurostat estimates.
(3) IT, LU, PT, UK and LI — 2001 population data: estimated values.
(4) IE and EE — 2001 population data: provisional value.

Sources: Eurostat, EPO.

In 2001, high tech patent applications to the EPO from the EU amounted to 11 928. Of these, 33.7% were applied for by German inventors, 17.9% by inventors from the UK and 15.0% by French inventors – Figure 5.16. Compared to the distribution of total patents, it may be seen that in high tech patents, the share accounted for by Germany reduces by 8.2%, whereas that of the United Kingdom and France increases by 4.8% and 0.9%, respectively.

As a proportion of population, the EU recorded an average of 32 high tech patent applications per million inhabitants. At the national level, the highest ratio was retained by Finland with 136 high tech patents per million inhabitants, followed by Sweden with 101 patent applications per million inhabitants – Figure 5.17. The Netherlands (69), Germany (49), Denmark (42) and the United Kingdom (36) also registered rates above the EU average.

As shown in Table 5.5., the number of high tech patent applications to the EPO from Acceding Countries is still relatively low. In 2001, Hungary was the Acceding Country with most patent applications in the high tech field (43). However, with 9 high tech patent applications per million inhabitants in 2001, Slovenia leads in relative terms with a rate in relation to population above those of Greece, Spain, Italy and Portugal.

Figure 5.16. **Distribution of high tech patent applications to the EPO EU-15 by Member State 2001**

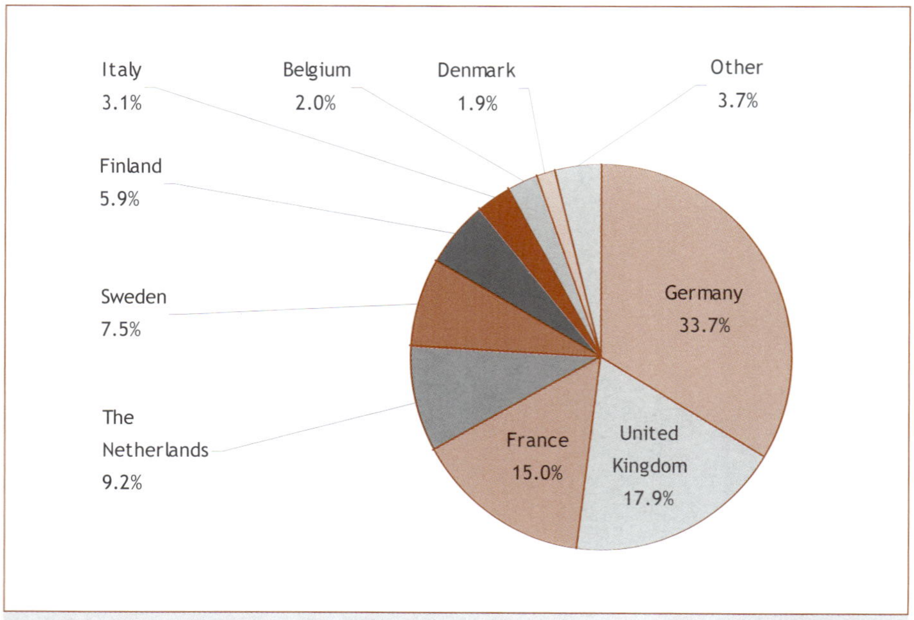

NB: Reference year corresponds to year of filing.
(1) 2001 provisional data.

Sources: Eurostat, EPO.

Figure 5.17. **Evolution of high tech patent applications to the EPO per million inhabitants EU-15, Iceland and Norway 1991, 1996 and 2001 (1)**

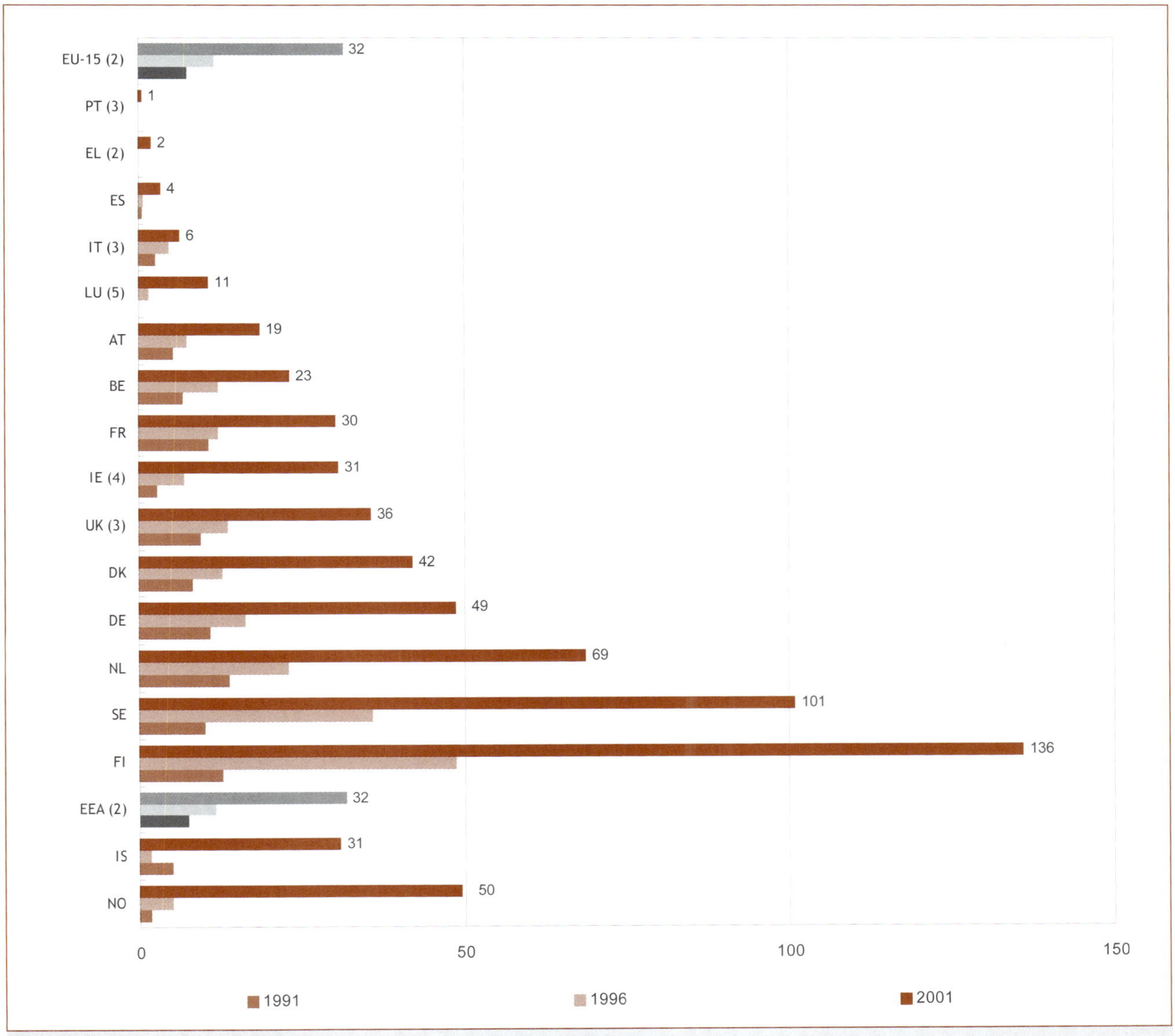

NB: Reference year corresponds to year of filing.

(1) 2001 provisional data.
(2) EU-15, EL and EEA — 2001 population data: Eurostat estimates.
(3) IT, PT and UK — 2001 population data: estimated values.
(4) IE — 2001 population data: provisional value.
(5) LU 1991: value equal to real zero.

Sources: Eurostat, EPO.

'Communication technology' accounts for almost half of all high tech patent applications from EU-15

Table 5.6. shows the distribution of high tech patent applications to the EPO by high tech group. As said earlier, high tech patents may be grouped in the following technology groups:

- Aviation — AVI,
- Computer and automated business equipment — CAB,
- Communication technology — CTE,
- Lasers — LSR,
- Micro-organism and genetic engineering — MGE,
- Semi-conductors — SMC.

Most high tech patent applications to the EPO from EU-15 in 2001 were done in the field of 'Communication technology' (47.1% of high tech applications), which includes electrical communication systems such as telephones or television. Together with CAB, they account for 75.6% of total applications. Among Member States, Finland appears highly specialised in this field, as 80.5% of the high tech applications from Finnish inventors were done in 'Communication technology'. To a lesser extent, also Sweden (60.5%) and Netherlands (52.8%) are specialised in 'Communication technology'. 'Computer and automated business equipment' was the most important high tech group for Greece (45.4%), whereas 'Micro-organism and genetic engineering' was the largest for Denmark (36.5%). Similar exceptions to the general pattern can be noticed for Luxembourg (52.2% in CAB) and Portugal (44.9% in MGE), but due to the low absolute figures they are less striking.

Figure 5.18. **Patent applications of 50 selected companies 1990 to 1997**

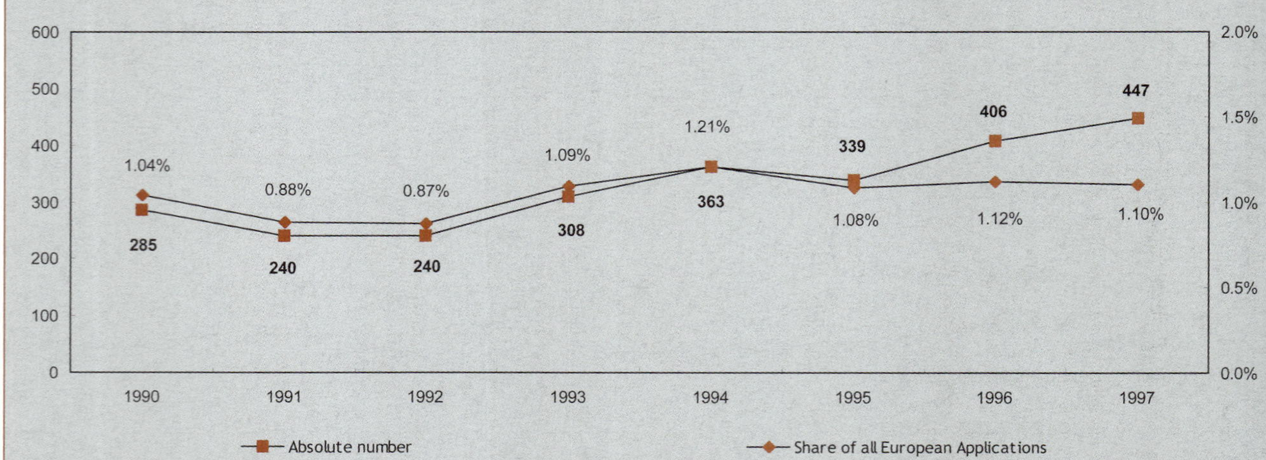

Source: *Third European Report on Science and Technology Indicators*, Directorate-General for Research, 2003, p. 409.
Data: EPO — data processed by Fraunhofer-ISI.

Patenting activities and innovation in the services sectors

Services encompass a broad spectrum of activities, ranging from retailing, architectural, software consulting, engineering and to public services such as the mail system and public transportation. There is no doubt about the fact that innovation exists in these sectors. Characteristics of service innovation still allow the use of the traditional taxonomy into product (service), process, organisational and market innovations. However, traditionally, a significant part of innovation patterns in services has been 'soft' or non technological. Innovation in services is especially present in knowledge-intensive services — KIS. However, innovation in services is still relatively under explored, although with the proliferating use of Information Technology in the delivery, use and composition of services, understanding how intellectual property rights are protected in the services sectors is becoming increasingly important.

In this context, a study which intends to find empirical evidence about patenting activities of services companies has been carried out by DG Research. The study selected 50 companies and then identified all the patent applications they made to the EPO between 1990 and 1997. Figure 5.18. shows the evolution of patent applications of 50 selected services companies. Going from 285 patent applications to 447, the number of patent applications to the EPO by this group of services firms almost doubled during the 1990-97 period. However, when compared to the total of patent applications, the percentage share of the sample group is just over 1%.

For further details on Patenting activities in the services sectors please refer to:
Third European Report on Science and Technology Indicators, Dossier V p. 407 ff, http://www.cordis.lu/indicators.

Overall, the 'Communication technology' field together with 'Computer and automated business equipment' account for most high tech patent applications from the Acceding Countries — 77.9% of total applications.

Table 5.6.

Distribution of high tech patent applications to the EPO, by high tech group EU-15, Candidates Countries, Iceland, Liechtenstein and Norway 2001 (1)

	Total number	Distribution by high tech group in % (2)					
		AVI	CAB	CTE	LSR	MGE	SMC
EU-15	11 928	1.2	28.5	47.1	1.4	13.0	8.9
EUR-12	8 673	1.4	26.2	47.4	1.3	12.9	10.9
BE	240	0.6	26.6	34.5	1.3	21.0	16.1
DK	225	0.2	26.8	33.6	0.9	36.5	2.0
DE	4 017	1.7	24.6	42.6	1.7	15.1	14.3
EL	22	2.3	45.4	29.0	-	9.8	13.6
ES	143	2.1	30.2	38.6	-	25.0	4.1
FR	1 791	2.3	31.4	46.0	1.4	11.8	7.1
IE	117	0.9	41.9	44.7	1.5	8.4	2.7
IT	374	0.5	30.0	43.4	2.5	13.1	10.6
LU	5	2.7	52.2	38.2	-	6.9	-
NL	1 100	0.2	26.7	52.8	0.1	9.3	10.8
AT	152	1.3	25.5	39.4	2.0	16.9	14.9
PT	7	-	21.7	19.1	-	44.9	14.3
FI	705	-	15.3	80.5	0.1	3.4	0.8
SE	896	0.9	26.9	60.5	1.2	6.9	3.5
UK	2 134	0.9	38.6	41.7	1.9	13.1	3.9
ACC	89	3.4	31.2	46.7	-	17.0	1.7
CZ	7	14.8	44.3	14.8	-	26.1	-
EE	2	-	50.0	-	-	50.0	-
CY	2	-	50.0	50.0	-	-	-
LV	1	100.0	-	-	-	-	-
LT	3	-	-	-	-	100.0	-
HU	43	-	35.3	55.2	-	7.3	2.3
MT	-	-	-	-	-	-	-
PL	8	13.1	30.5	32.8	-	23.6	-
SI	17	-	23.1	57.7	-	19.2	-
SK	6	-	16.9	50.6	-	24.0	8.5
BG	3	-	66.9	-	-	33.1	-
RO	3	66.7	-	-	-	-	33.3
TR	12	-	41.7	50.0	-	8.3	-
EEA	12 160	1.2	28.7	47.0	1.4	13.0	8.7
IS	9	-	22.6	28.4	-	37.5	11.4
LI	-	-	-	-	-	-	-
NO	223	1.1	38.8	43.0	1.8	14.3	0.9

NB: Reference year corresponds to year of filing.

(1) 2001 provisional data.
(2) See abbreviations on page 168.

Sources: Eurostat, EPO.

5.4. Performance at the regional level in the EEA

This section provides an insight into the regional patenting activities. The analysis covers the EEA regions with special focus on EU-15. Regions are considered at the NUTS 2 level. For Denmark and Luxembourg, the entire national territory is classified as a NUTS 2 region, which explains their potential appearance in the regional rankings.

Total patent applications to the EPO

Île de France (FR) leads in absolute terms and Oberbayern (DE) as a proportion of population

Map 5.1. gives an overview of the inventive performance of the EEA regions, as it depicts the ratios of patent applications to the EPO per million inhabitants.

Patent applications to the EPO per million inhabitants in the EEA ranged from 824 in Oberbayern (DE) to zero applications in various regions of Greece (Dytiki Makedonia, Ipeiros, Anatoliki Makedonia-Thraki, Thessalia, Ionia Nisia, Voreio Aigaio), Spain (Ceuta y Melilla), France (Martinique, French Guiana) and Portugal (Acores, Madeira). Three main areas where regions with high levels of patent applications per million inhabitants are concentrated: regions in Northern Europe, in central Europe — particularly the regions surrounding Bayern and Baden-Württemberg (DE) — and the South of the United Kingdom. In addition, Île de France and Rhône-Alpes in France also feature among the leading EU regions.

Map 5.1.

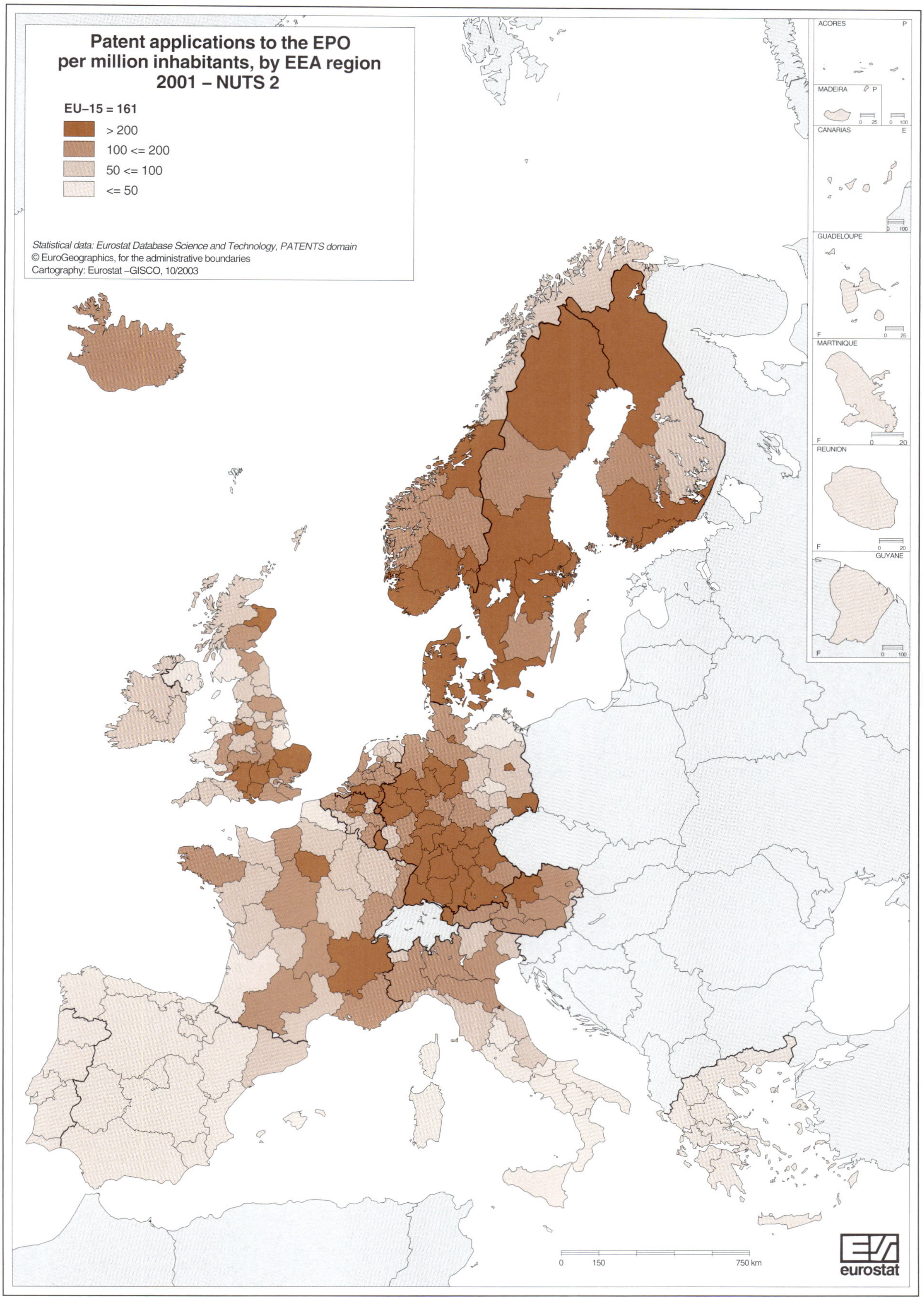

NB: Reference year corresponds to year of filing.

EU-15 = 161 refers to the EU-15 average, i.e. in 2001 inventors from the EU applied at the EPO for 161 patents per million inhabitants.
2001 provisional data;
All regions of EU-15 except for those in ES, LU and DK — 2001 population data: Eurostat estimates;
LU and LI — 2001 population data: estimated values;
EU-15 — 2001 population data: Eurostat estimates.

As shown in Table 5.7., the EU region that recorded the highest ratio of patent applications to the EPO per million inhabitants in 2001 was Oberbayern (824, DE), followed by the Dutch region of Noord-Brabant (822) and Stuttgart in Germany (719).

In absolute terms, Table 5.8. illustrates that Île de France was leading, with 3 423 patent applications, ahead of Oberbayern (3 325) and Stuttgart (2 817). Out of the 211 EU regions at the NUTS 2 level, the top 15 regions ranked in Table 5.8. account for 42.9% of the total number of patent applications from the EU, showing thus a very high concentration of patenting activities.

The dominance of German regions among the leaders in patenting may be observed as 9 of the 15 leading regions both in absolute and relative terms belonged to this country. Two regions from Sweden and one region from the Netherlands, Finland, Austria and Belgium were also present in the top 15 as a proportion of population. In absolute terms, besides the 9 German regions, there were 2 regions from France and one from the Netherlands, Italy, Denmark and Sweden.

Table 5.9. shows the top three patenting regions of each Member State in absolute terms in 2001. It provides details on the total number of applications to the EPO, their ratio per million inhabitants and their corresponding distribution by IPC section.

The leading regions in absolute terms for each Member State were: Antwerp in Belgium, Oberbayern in Germany, Attiki in Greece, Cataluña in Spain, Île de France in France, Southern and Eastern in Ireland, Lombardia in Italy, Noord-Brabant in the Netherlands, Oberösterreich in Austria, Lisboa e Vale do Tejo in Portugal, Uusimaa (Suuralue) in Finland, Stockholm in Sweden and East Anglia in the United Kingdom.

Concerning the distribution by IPC section, the percentage of patent applications accounted for by each IPC section varies across countries and regions. However, leading regions are often specialised in 'Electricity – Section H' and in 'Performing operations; transporting – Section B', being therefore in line with the distribution for the EU average. The specialisation in 'Electricity – Section H' is most evident for the leading regions of Finland, Sweden and the United Kingdom.

The lead of German regions in absolute terms is also evident as regards to Table 5.9., as even the third region in this country is above the top regions of the rest of the Member States – with the exception of the French capital region of Île de France and the Dutch region of Noord-Brabant.

Table 5.7. Top fifteen patenting regions in terms of applications per million inhabitants EU-15 2001 (1)

Ranking	Country	NUTS 2 region	Patent applications: Per million inhabitants	Patent applications: Total number
1	DE	Oberbayern	824	3 325
2	NL	Noord-Brabant	822	1 937
3	DE	Stuttgart	719	2 817
4	SE	Stockholm	610	1 101
5	FI	Uusimaa (Suuralue)	582	803
6	DE	Mittelfranken	518	872
7	DE	Rheinhessen-Pfalz	494	990
8	DE	Karlsruhe	493	1 319
9	DE	Darmstadt	491	1 825
10	DE	Tübingen	481	845
11	DE	Freiburg	474	1 008
12	AT	Vorarlberg	453	158
13	BE	Brabant Wallon	448	157
14	SE	Sydsverige	435	555
15	DE	Köln	395	1 684
		EU-15	161	60 890

NB: Reference year corresponds to year of filing.

(1) 2001 provisional data.
All regions of EU-15 **except for those in DK, ES and LU** —
2001 population data: Eurostat estimates;
LU — 2001 population data: estimated values;
EU-15 — 2001 population data: Eurostat estimates.

Sources: Eurostat, EPO.

Table 5.8. Top fifteen patenting regions in terms of total number of applications EU-15 2001 (1)

Ranking	Country	NUTS 2 region	Patent applications: Total number	Patent applications: Per million inhabitants
1	FR	Île de France	3 423	312
2	DE	Oberbayern	3 325	824
3	DE	Stuttgart	2 817	719
4	NL	Noord-Brabant	1 937	822
5	DE	Darmstadt	1 825	491
6	DE	Düsseldorf	1 788	340
7	DE	Köln	1 684	395
8	IT	Lombardia	1 528	169
9	FR	Rhône-Alpes	1 383	244
10	DE	Karlsruhe	1 319	493
11	DK	Denmark	1 129	211
12	SE	Stockholm	1 101	610
13	DE	Freiburg	1 008	474
14	DE	Rheinhessen-Pfalz	990	494
15	DE	Mittelfranken	872	518
		EU-15	60 890	161

NB: Reference year corresponds to year of filing.

(1) 2001 provisional data.
All regions of EU-15 **except for those in DK, ES and LU** —
2001 population data: Eurostat estimates;
LU — 2001 population data: estimated values;
EU-15 — 2001 population data: Eurostat estimates.

Sources: Eurostat, EPO.

Table 5.9.

Top three patenting regions in terms of total number of applications EU-15 by Member State 2001 (1, 2)

EU-15 Ranking in relative terms		NUTS 2 region	Total number	Per million inhabitants	Distribution by IPC section in % (3)							
					A	B	C	D	E	F	G	H
		EU-15	**60 890**	**161**	**15.0**	**19.4**	**14.3**	**1.9**	**4.2**	**9.8**	**16.7**	**18.8**
50	BE	Antwerpen	334	203	11.8	22.6	21.0	1.1	2.5	6.2	20.7	14.1
35	BE	Vlaams Brabant	242	238	13.7	12.7	37.4	1.8	2.5	1.8	13.5	16.6
83	BE	Oost-Vlaanderen	174	127	20.9	13.2	22.3	2.3	3.1	3.8	16.8	17.7
46	DK	Denmark	1 129	212	24.1	13.0	18.4	1.1	4.6	8.7	16.1	14.1
1	DE	Oberbayern	3 325	824	8.6	15.8	7.6	0.5	2.0	9.2	23.8	32.3
3	DE	Stuttgart	2 817	719	4.0	25.7	3.5	1.8	3.4	27.3	16.3	18.1
9	DE	Darmstadt	1 825	491	16.8	21.3	29.3	1.8	2.5	7.8	9.9	10.6
172	EL	Attiki	44	13	18.9	24.0	15.6	-	8.0	3.4	8.5	21.5
178	EL	Kentriki Makedonia	18	10	22.5	-	0.7	-	-	22.5	34.5	19.9
174	EL	Kriti	7	12	53.8	15.1	11.1	-	-	-	-	20.0
142	ES	Cataluna	382	62	23.9	26.2	13.2	4.0	4.7	5.5	10.9	11.6
162	ES	Comunidad de Madrid	187	36	19.0	16.3	19.0	-	3.5	4.5	16.7	20.9
167	ES	Comunidad Valenciana	104	26	23.6	15.4	24.3	2.4	10.1	5.2	11.9	7.1
26	FR	Île de France	3 423	312	18.8	15.7	12.1	0.3	2.9	9.7	19.3	21.3
31	FR	Rhône-Alpes	1 383	244	18.9	16.5	18.6	2.4	3.0	6.6	15.9	18.2
92	FR	Provence-Alpes-Côte d'Azur	516	114	15.8	11.0	12.4	0.1	2.5	3.3	35.2	19.7
107	IE	Southern and Eastern	262	94	17.7	12.1	10.7	0.1	4.5	4.1	29.9	20.8
135	IE	Border, Midlands and Western	65	66	36.9	18.7	0.8	0.5	1.5	1.9	9.8	29.9
59	IT	Lombardia	1 528	169	20.1	22.4	14.3	4.8	4.8	9.6	10.1	13.9
56	IT	Emilia-Romagna	703	177	19.1	45.0	7.5	1.1	5.9	11.3	6.0	4.1
95	IT	Veneto	496	110	29.3	22.7	7.6	3.1	7.3	12.0	8.8	9.1
44	LU	Luxembourg	93	211	2.8	28.4	21.5	-	9.1	21.4	8.7	8.0
2	NL	Noord-Brabant	1 937	822	6.0	6.0	4.3	0.4	0.6	2.6	32.4	47.7
69	NL	Zuid-Holland	509	150	26.6	14.0	22.3	1.1	9.3	4.6	14.4	7.7
74	NL	Noord-Holland	354	140	16.3	22.7	27.9	0.7	5.3	4.9	14.3	7.8
49	AT	Oberösterreich	283	205	7.8	32.7	17.9	3.7	9.2	17.4	6.3	5.1
67	AT	Wien	251	156	19.7	15.2	14.9	0.7	4.9	2.1	21.2	21.1
53	AT	Steiermark	221	184	7.5	26.3	19.2	3.6	8.0	9.0	10.2	16.2
186	PT	Lisboa e Vale do Tejo	23	7	41.2	7.4	25.3	-	3.3	10.3	2.2	10.2
188	PT	Norte	21	6	19.4	34.8	26.7	3.1	-	5.1	5.3	5.7
187	PT	Centro (PT)	11	6	1.5	11.8	11.9	-	35.8	29.9	9.0	-
5	FI	Uusimaa (Suuralue)	803	582	8.7	10.7	8.7	7.7	1.6	2.8	15.1	44.7
23	FI	Etelä-Suomi	597	328	6.7	18.6	6.1	6.8	3.2	5.4	17.4	35.8
25	FI	Pohjois-Suomi	180	323	8.4	10.2	2.7	1.7	1.3	3.6	12.9	59.2
4	SE	Stockholm	1 101	610	14.3	9.8	7.2	1.1	1.9	5.6	18.9	41.1
21	SE	Västsverige	605	343	22.6	23.3	6.9	4.7	3.2	13.0	12.4	13.9
14	SE	Sydsverige	555	435	15.4	18.4	8.0	1.4	3.2	7.4	22.1	24.1
20	UK	East Anglia	784	356	11.7	9.0	14.1	0.4	2.9	3.4	26.2	32.3
18	UK	Berks., Bucks and Oxfords.	764	360	17.6	10.6	19.4	0.6	1.6	4.4	28.7	17.2
34	UK	Gloucesters., Wilts. & North Somerset	522	239	12.9	12.6	5.7	2.3	3.5	4.5	28.0	30.5

NB: Reference year corresponds to year of filing.

(1) 2001 provisional data.

(2) All regions of EU-15 **except for those in DK, ES and LU —** 2001 population data: Eurostat estimates;
LU — 2001 population data: estimated values;
EU-15 — 2001 population data: Eurostat estimates.

(3) A Human necessities;
B Performing operations; transporting;
C Chemistry; metallurgy;
D Textiles; paper;
E Fixed constructions;
F Mechanical engineering; lighting; heating; weapons; blasting;
G Physics;
H Electricity.

Sources: Eurostat, EPO.

Map 5.2.

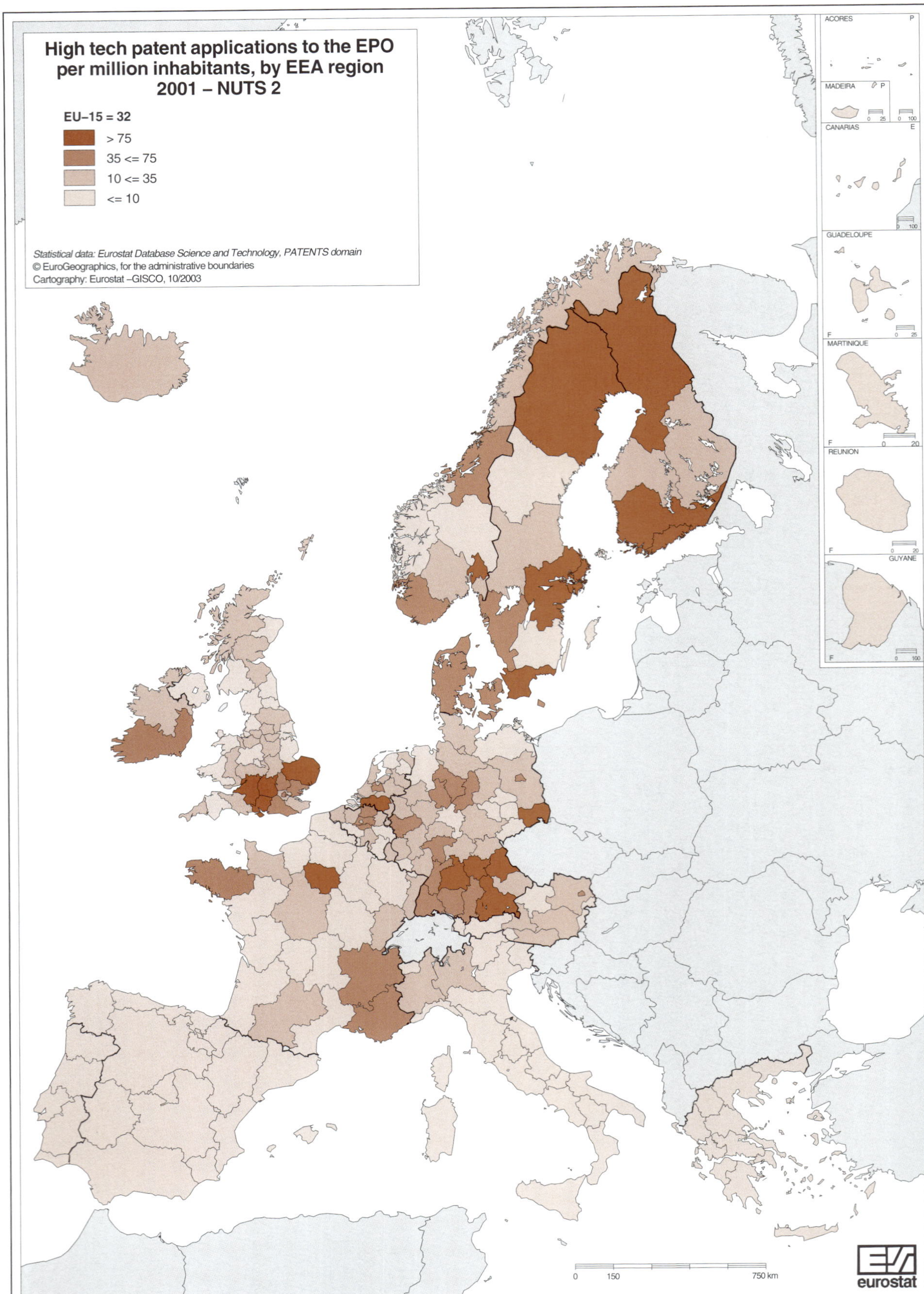

NB: Reference year corresponds to year of filing.

EU-15 = 32 refers to the EU-15 average, i.e. in 2001 inventors from the EU applied at the EPO for 32 high tech patents per million inhabitants.
2001 provisional data;
All regions of EU-15 **except for those in DK, ES and LU** — 2001 population data: Eurostat estimates.
LU and LI — 2001 population data: estimated values.
EU-15 — 2001 population data: Eurostat estimates.

High tech patent applications to the EPO

Oberbayern (DE) leads in absolute terms, as does Noord-Brabant (NL) relative to population

Map 5.2. shows the regional distribution of high tech patenting in the EU at the NUTS 2 level. In 2001, high tech patent applications per million inhabitants in the EEA regions ranged between 342 in Noord-Brabant (NL) to zero applications in various regions of Greece (Dytiki Makedonia, Ipeiros, Anatoliki Makedonia-Thraki, Thessalia, Ionia Nisia, Sterea Ellada, Voreio Aigaio), Spain (Cantabria, La Rioja, Castilla-la Mancha, Extremadura, Ceuta y Melilla), France (Corse, Martinique, Guadeloupe, French Guiana, Reunion) and Portugal (Alentejo, Algarve, Acores, Madeira).

Relative to the population, the EU region that registered the highest rate in 2001 was Noord-Brabant in the Netherlands, as it recorded a rate of 342 high tech patent applications per million inhabitants — Table 5.10. Following Noord-Brabant were the Finish region of Uusimaa (286) and Oberbayern (282) in Germany.

The top fifteen regions in terms of the total number of patent applications to the EPO in 2001 are listed in Table 5.11. The EU region that applied for most high tech patents at the EPO was Oberbayern in Germany — 1 138 high tech patent applications, followed by Île de France (886), the Dutch region of Noord-Brabant (805), Stockholm in Sweden (444) and Uusimaa in Finland (395).

It may be observed that the inventive activity in the high tech fields is more spread across the Member States of the EU and is less concentrated in German regions as it is the case for total patenting. The United Kingdom retained the largest number of regions in the top 15: 4 regions in the ranking as a proportion of population and 5 in the ranking in absolute terms.

Table 5.10. Top fifteen high tech patenting regions in terms of applications per million inhabitants EU-15 2001 (1)

			High tech patent applications	
Ranking	Country	NUTS 2 region	Per million inhabitants	Total number
1	NL	Noord-Brabant	342	805
2	FI	Uusimaa (Suuralue)	286	395
3	DE	Oberbayern	282	1 138
4	SE	Stockholm	246	444
5	UK	East Anglia	168	369
6	FI	Pohjois-Suomi	151	84
7	UK	Hamps. & Isle of Wight	145	258
8	SE	Sydsverige	142	181
9	FI	Etelä-Suomi	112	204
10	DE	Mittelfranken	104	175
11	UK	Berks., Bucks & Oxfords.	101	214
12	DE	Stuttgart	95	371
13	UK	Gloucesters., Wilts. & North Somerset	94	206
14	FR	Île de France	81	886
15	SE	Östra Mellansverige	80	119
		EU-15	**32**	**11 928**

NB: Reference year corresponds to year of filing.

(1) 2001 provisional data.
All regions of EU-15 **except for those in DK, ES and LU —** 2001 population data: Eurostat estimates;
LU — 2001 population data: estimated values;
EU-15 — 2001 population data: Eurostat estimates.

Sources: Eurostat, EPO.

Table 5.11. Top fifteen high tech patenting regions in terms of total number of applications EU-15 2001 (1)

			High tech patent applications	
Ranking	Country	NUTS 2 region	Total number	Per million inhabitants
1	DE	Oberbayern	1 138	282
2	FR	Ile de France	886	81
3	NL	Noord-Brabant	805	342
4	SE	Stockholm	444	246
5	FI	Uusimaa (Suuralue)	395	286
6	DE	Stuttgart	371	95
7	UK	East Anglia	369	168
8	UK	Hamps. & Isle of Wight	258	145
9	FR	Rhône-Alpes	240	42
10	DE	Köln	233	55
11	DK	Denmark	225	42
12	UK	Berks., Bucks & Oxfords.	214	101
13	UK	Gloucesters., Wilts. & North Somerset	206	94
14	FI	Etelä-Suomi	204	112
15	UK	Inner London	202	71
		EU-15	**11 928**	**32**

NB: Reference year corresponds to year of filing.

(1) 2001 provisional data.
All regions of EU-15 **except for those in DK, ES and LU —** 2001 population data: Eurostat estimates;
LU — 2001 population data: estimated values;
EU-15 — 2001 population data: Eurostat estimates.

Sources: Eurostat, EPO.

Table 5.12. **Top three high tech patenting regions in terms of total number of applications EU-15 by Member State 2001 (1, 2)**

EU-15 ranking in relative terms	Country	NUTS 2 region	Total number	Per million inhabitants	Distribution by high tech group in % (2)					
					AVI	CAB	CTE	LSR	MGE	SMC
		EU-15	11 928	32	1.2	28.5	47.1	1.4	13.0	8.9
40	BE	Antwerpen	62	38	1.6	37.2	45.4	0.4	7.1	8.2
26		Vlaams Brabant	51	50	-	17.6	27.7	-	23.1	31.6
50		Oost-Vlaanderen	41	30	-	14.2	38.7	6.1	25.1	15.9
35	DK	Denmark	225	42	0.2	26.8	33.6	0.9	36.5	2.0
3	DE	Oberbayern	1 138	282	1.3	26.5	47.8	1.6	6.2	16.5
12		Stuttgart	371	95	1.1	28.4	52.2	3.7	5.5	9.2
22		Köln	233	55	0.5	23.8	44.3	-	23.3	8.2
157	EL	Attiki	11	3	4.6	21.5	40.3	-	9.0	24.5
155		Kentriki Makedonia	6	3	-	83.3	16.7	-	-	-
153		Dytiki Ellada	3	4	-	100.0	-	-	-	-
116	ES	Cataluna	51	8	1.0	57.0	21.1	-	13.6	7.2
108		Comunidad de Madrid	47	9	5.3	9.2	59.7	-	21.1	4.7
148		Comunidad Valenciana	18	4	-	22.7	31.4	-	45.9	-
14	FR	Île de France	886	81	1.6	27.8	53.4	2.2	10.9	4.2
34		Rhône-Alpes	240	42	-	35.6	29.0	0.6	14.4	20.3
32		Provence-Alpes-Côte d'Azur	199	44	2.2	59.0	28.5	0.2	5.1	5.0
41	IE	Southern and Eastern	101	36	1.0	48.5	35.9	1.7	9.7	3.1
82		Border, Midlands and Western	16	16	-	-	100.0	-	-	-
75	IT	Lombardia	174	19	1.1	27.7	47.8	1.3	11.9	10.1
103		Piemonte	43	10	-	38.5	42.9	6.1	1.2	11.3
130		Lazio	31	6	-	16.2	38.0	-	34.6	11.2
98	LU	Luxembourg	5	11	2.7	52.2	38.2	-	6.9	-
1	NL	Noord-Brabant	805	342	-	27.6	57.7	0.1	1.0	13.5
65		Zuid-Holland	76	22	2.2	25.1	30.3	-	38.4	4.1
62		Noord-Holland	60	24	-	33.1	31.2	-	35.7	-
38	AT	Wien	66	41	-	22.7	55.5	1.5	19.6	0.7
71		Niederösterreich	31	20	-	25.2	50.1	1.6	20.2	2.9
95		Steiermark	14	12	-	32.5	8.8	-	29.3	29.3
171	PT	Lisboa e Vale do Tejo	5	1	-	10.6	28.1	-	40.2	21.1
182		Norte	2	1	-	48.8	-	-	51.2	-
188		Centro (PT)	0	0	-	-	-	-	100.0	-
2	FI	Uusimaa (Suuralue)	395	286	-	15.4	80.0	-	3.5	1.1
9		Etelä-Suomi	204	112	-	16.2	80.0	0.3	3.1	0.5
6		Pohjois-Suomi	84	151	-	11.2	86.2	-	2.5	-
4	SE	Stockholm	444	246	0.3	21.6	68.1	1.6	5.1	3.3
8		Sydsverige	181	142	0.6	41.1	50.9	0.2	7.0	0.3
15		Östra Mellansverige	119	80	3.4	29.7	44.1	-	16.3	6.5
5	UK	East Anglia	369	168	0.1	31.0	45.6	3.8	13.2	6.3
7		Hamps. & Isle of Wight	258	145	-	43.1	48.4	2.6	1.2	4.7
11		Berks., Bucks & Oxfords.	214	101	1.5	47.2	36.1	0.6	12.6	2.0

NB: Reference year corresponds to year of filing.
Total figures are rounded, while percentages are calculated on precise total figures as resulting from fractional counting. For example, the real values for Centro (PT) are 0.17 high tech patent applications in 2001, corresponding to a ratio of 0.096 per million inhabitants.

(1) 2001 provisional data.
All regions of EU-15 **except for those in DK, ES and LU** — 2001 population data: Eurostat estimates;
LU 2001 population data: estimated values;
EU-15 2001 population data: Eurostat estimates.

(2) See abbreviations on page 168.

Sources: Eurostat, EPO.

Details for the top three regions of each Member State in terms of the total number of high tech patent applications to the EPO are provided in Table 5.12. Information is given on the total number of high tech applications to the EPO, their ratio per million inhabitants and their corresponding distribution by high tech group.

The leading regions in high tech patents in absolute terms for each Member State were: Antwerp in Belgium, Oberbayern in Germany, Attiki in Greece, Cataluña in Spain, Île de France in France, Southern and Eastern in Ireland, Lombardia in Italy, Noord-Brabant in the Netherlands, Wien in Austria, Lisboa e Vale do Tejo in Portugal, Uusimaa (Suuralue) in Finland, Stockholm in Sweden and East Anglia in the United Kingdom. With the exception of Austria, the leading region by Member State in high tech patenting coincides with those of patenting overall shown in Table 5.9.

The dominance of German regions is less striking in the high tech fields. As shown in Table 5.12., the leading region of France, (Île de France), the Netherlands (Noord-Brabant), Finland (Uusimaa) and Sweden (Stockholm) applied for more patents than the second German region.

Many of the leading regions by Member State are also specialised, as the EU average is, in 'Communication technology – CTE'. A very high proportion of patent applications in the 'Communication technology' field is retained by the top three leading regions of Finland: 80.0% in Uusimaa (Suuralue) and Etelä-Suomi and 86.2% in Pohjois-Suomi. The 'Computer and automated business equipment – CAB' and the 'Micro-organism and genetic engineering – MGE' fields were also the largest for several of the leading regions. All leading Portuguese regions were specialised in the field of the 'Micro-organism and genetic engineering'. Concerning the dynamics of high tech patent applications to the EPO at the regional level, an overall upward trend may be observed.

Table 5.13. shows the ten regions with the highest growth and those with the lowest growth when taking into consideration only regions that in 2001 registered a rate per million inhabitants above the EU average.

Among these regions, the region that recorded the highest annual average growth rate of high tech patent applications during the 1996-2001 period was Västsverige in Sweden (50.4% per annum), whereas the region of the Belgian capital city of Brussels (Région Bruxelles-capitale) grew at the slowest rate (5.5% per annum).

Table 5.13.

Regions with highest and lowest growth (1) in high tech patenting
EU-15
1996 to 2001 (2)

Regions with highest growth				Regions with lowest growth			
Country	NUTS 2 region	Total number 2001	AAGR (3) 1996-2001 in %	Country	NUTS 2 region	Total number 2001	AAGR (3) 1996-2001 in %
SE	Västsverige	85	50.4	BE	Région Bruxelles-capitale	30	5.5
DE	Detmold	106	49.5	BE	Antwerpen	62	5.8
DE	Braunschweig	70	40.0	UK	Surrey, East and West Sussex	103	8.6
FR	Bretagne	158	36.6	DE	Freiburg	98	9.6
NL	Noord-Brabant	805	33.7	FI	Pohjois-Suomi	84	11.3
IE	Southern and Eastern	101	33.6	UK	Bedfordshire, Hertfordshire	74	13.0
UK	Hampshire and Isle of Wight	258	33.4	NL	Utrecht	39	13.7
SE	Övre Norrland	40	31.3	AT	Wien	66	14.1
FR	Provence-Alpes-Côte d'Azur	199	31.2	BE	Vlaams Brabant	51	14.5
UK	Inner London	202	30.0	DE	Hannover	105	16.2

NB: Reference year corresponds to year of filing.
High tech patent applications in the EU grew during the 1996-2001 period at an annual average growth rate of 22.3%.

(1) With a ratio of high tech patent applications per million inhabitants at least equal to the EU average (32).
(2) 2001 provisional data.
(3) AAGR: Annual average growth rate.

Sources: Eurostat, EPO.

6.1. Introduction

Creating, exploiting and commercialising new technologies is vital for the competitiveness of a country in the modern economy. This is because high technology sectors are key drivers for economic growth, productivity and welfare, and are generally a source of high value added and well-paid employment.

The firms which are technology-intensive are known as high technology – or high tech – firms. These firms are vital to the competitiveness position of nations because:

- they are associated with innovation and hence tend to gain a larger market share, create new product markets, and use resources more productively
- they are linked to high value-added production and success in foreign markets, which sometimes helps to support higher returns to the workers they employ
- the industrial R&D they perform has spillover effects which benefit other commercial sectors by generating new products and processes, often leading to productivity gains, business expansions, and the creation of high wage jobs.

In this context, this chapter analyses Europe's performance in high technology sectors by looking at statistics on employment, value added and labour productivity in high tech and knowledge-intensive industries as well as international trade of high technology products. In order to perceive how European countries perform in comparison to their main competitors, other leading economies are also considered whenever possible.

- Firstly, in Section 6.2. the chapter looks at the evolution and distribution of employment in high tech and knowledge-intensive sectors both at the national and regional levels. Covering the period 1997-2002, national data are given for the 15 Member States of the EU and the Candidate Countries. Regional data are analysed at the NUTS 2 level covering EU-15, Iceland and Norway. Employment in high tech and knowledge-intensive services are extracted and built up using data from the *EU Labour Force Survey* – EU LFS.

- Secondly, in Section 6.3. an overview is provided on statistics on value added and labour productivity by looking at the performance on selected sectors in 2000:
  - high tech manufacturing,
  - medium-high tech manufacturing,
  - knowledge-intensive market services and
  - high tech services.

  Here, EU and Candidate Countries are considered at the national level. These data have been obtained from ***Structural Business Statistics*** – SBS – database.

- Finally, in Section 6.4. the analysis describes the evolution of international high tech trade which makes up a considerable proportion of total trade in many advanced economies. The data generally cover the reference period 1996-2001 and international comparison is made between the EU, Japan and the United States. Where the relevant data are available, the Acceding Countries aggregate – ACC – is also considered. EU-15 aggregate data refer to extra EU trade, i.e. they exclude trade within the countries of the European Union. Data for individual Member States include both intra and extra EU trade, unless otherwise stated.

  All high tech trade data relating to the EU countries are based on data extracted from the *COMEXT* database – Eurostat's database of official statistics on EU's external trade and trade between EU Member States. This database includes imports and exports data flows with Member States and third countries as reported by the EU countries only. Trade data reported by third countries – including the Acceding Countries were therefore extracted from the UN statistical office's *Comtrade* database.

  For a detailed definition of high tech products and sectors please refer to the methodological notes starting on page 150.

6.2. Employment in high tech and knowledge-intensive sectors in the EU and Candidate Countries

At the national level

In the EU, 33.3% of workers are employed in KIS and 7.4% in high tech and medium-high tech manufacturing

Figure 6.1. shows the distribution of employment in the EU and Candidate Countries by selected sectors in 2002. With almost 163 million people employed in the EU, services sectors accounted for 68% of total employment in 2002, among which knowledge-intensive services – KIS – are becoming increasingly important: 33.3% of total employment. Whilst high tech and medium-high tech manufacturing sectors account for 7.4% of employment, other manufacturing sectors employ 11.8% of the EU's workforce and other non manufacturing and non services sectors – i.e. agriculture, fishing, mining, construction, etc. – 12.9%.

At the Member State level, Germany was the country where high tech and medium-high tech manufacturing sectors accounted for the largest proportion, 11.4% of total employment, in 2002. Following Germany were Italy and Finland, both 7.4%. The rest of the EU Member States recorded rates that were below the EU average.

Figure 6.1.

Distribution of employment by selected sector EU-15 and Candidate Countries (1) 2002

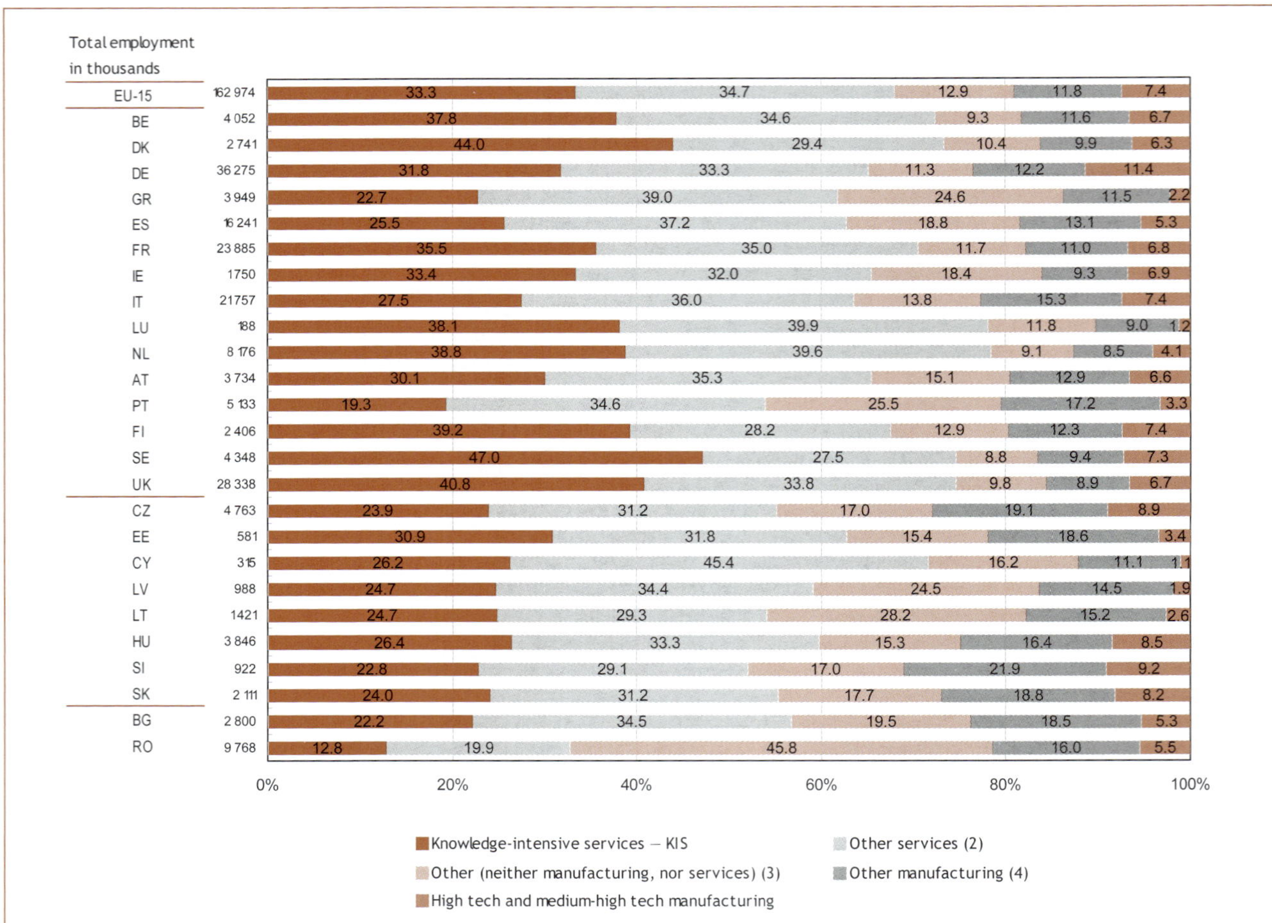

(1) Data for MT, PL and TR are not available, as there are not EU LFS data with the necessary breakdowns to construct high tech and knowledge-intensive employment indicators.
(2) 'Other services' refers to total services excluding knowledge-intensive services — KIS.
(3) 'Other (neither manufacturing, nor services)' refers to total economy excluding manufacturing and services sectors.
(4) 'Other manufacturing' refers to total manufacturing excluding high tech and medium-high tech manufacturing sectors.

Source: Eurostat, EU LFS — spring data.

Concerning the proportion of employment accounted for by knowledge-intensive services , in 2002, Sweden was the EU Member State most specialised in these sectors: 47.0% of total employment. With the exception of Germany, Greece, Spain, Italy, Austria and Portugal, employment in KIS accounted for a proportion above the EU average in all other Member States of the Union.

As for the Acceding Countries, Slovenia recorded the highest percentage of people employed in high tech and medium-high tech manufacturing sectors (9.2%). The Czech Republic (8.9%), Hungary (8.5%) and Slovak Republic (8.2%) also recorded rates above the EU average. As regards knowledge-intensive services, Estonia had the highest percentage of people employed in KIS (30.9%), it remains however below the EU mean.

Looking at the distribution of employment by gender, female employment appears even more skewed towards knowledge-intensive services, as 45.2% of the EU's female employment was in these sectors — Table 6.1. On the contrary, employment in knowledge-intensive services only accounted for 24.3% of male employment in the EU. Male employment seems to be more specialised in high tech and medium-high tech manufacturing sectors. These sectors accounted for 9.8% of total male employment, whereas they only represented 4.1% of the EU's female employment.

Table 6.1. **Distribution of employment by selected sector and gender**
EU-15 and Candidate Countries
2002

	Women					Men				
	KIS	Other services (1)	Other (neither manufacturing nor services) (2)	Other manufacturing (3)	High tech and medium-high tech manufacturing	KIS	Other services (1)	Other (neither manufacturing nor services) (2)	Other manufacturing (3)	High tech and medium-high tech manufacturing
EU-15	45.2	37.1	5.1	8.5	4.1	24.3	32.8	18.8	14.3	9.8
BE	50.8	35.6	2.5	7.2	3.9	28.2	33.9	14.3	14.9	8.8
DK	58.8	27.3	3.0	6.7	4.3	31.0	31.2	17.0	12.7	8.1
DE	42.8	37.6	4.7	8.9	6.0	22.9	29.9	16.6	14.8	15.6
EL	30.8	40.2	18.6	9.3	1.2	17.7	38.3	28.3	12.9	2.8
ES	37.4	43.9	6.1	9.5	3.2	18.4	33.2	26.4	15.3	6.7
FR	46.6	37.2	4.6	7.4	4.2	26.3	33.1	17.6	14.0	9.0
IE	47.9	36.9	3.2	5.8	6.2	22.9	28.5	29.3	11.9	7.4
IT	39.1	36.9	5.7	13.9	4.4	20.4	35.5	18.8	16.2	9.1
LU	51.7	40.7	2.8	3.8	1.1 u	28.9	39.4	17.8	12.5	1.3
NL	50.9	39.2	3.2	5.0	1.6	29.4	39.8	13.6	11.3	6.0
AT	40.6	40.4	8.1	7.6	3.4	21.5	31.3	20.8	17.3	9.2
PT	27.7	37.1	15.3	17.2	2.8	12.4	32.5	34.1	17.3	3.8
FI	53.4	29.2	5.3	8.3	3.8	26.0	27.2	20.0	16.1	10.7
SE	62.6	25.9	2.5	5.2	3.9	32.8	29.0	14.6	13.3	10.4
UK	52.7	35.8	2.7	5.4	3.4	31.0	32.1	15.7	11.8	9.5
CZ	35.6	33.2	6.7	17.3	7.3	14.8	29.6	24.9	20.4	10.2
EE	40.9	31.8	6.0	17.3	4.1	21.1	31.7	24.5	19.9	2.7
CY	34.1	49.5	6.1	9.1	1.2	19.9	42.2	24.2	12.7	1.0
LV	35.2	37.5	14.0	12.4	1.0	14.5	31.5	34.6	16.6	2.8
LT	36.4	29.1	16.8	16.0	1.8	13.5	29.4	39.2	14.4	3.5
HU	38.1	33.9	5.8	15.5	6.8	16.9	32.9	23.2	17.1	9.9
SI	31.1	30.8	11.2	19.1	7.7	15.8	27.6	21.9	24.2	10.5
SK	35.6	33.9	6.5	17.4	6.6	14.3	29.0	27.2	20.0	9.5
BG	30.9	32.2	11.4	21.6	3.9	14.3	36.6	26.8	15.7	6.6
RO	17.7	17.8	42.4	18.2	3.9	8.6	21.7	48.7	14.2	6.9

(1) 'Other services' refers to total services excluding knowledge-intensive services — KIS.
(2) 'Other (neither manufacturing, nor services)' refers to total economy excluding manufacturing and services sectors.
(3) 'Other manufacturing' refers to total manufacturing excluding high tech and medium-high tech manufacturing sectors.

Source: Eurostat, EU LFS — spring data.

Whilst Ireland is most specialised in employment in high tech manufacturing (3.2% of employment), Germany is in employment in medium-high tech (9.4%)

Looking at high tech and medium-high tech manufacturing sectors alone, in 2002 there were 12 million people employed in these sectors in the EU, of which over 2 million were working in high tech manufacturing sectors – Table 6.2. The EU Member State with most people employed in high tech and medium-high tech manufacturing in 2002 was Germany (4 122 thousand), followed by the United Kingdom (1 901 thousand), France (1 628 thousand) and Italy (1 603 thousand). Among the Acceding Countries, the Czech Republic registered the largest number of people employed in these sectors (425 thousand). The percentage of women in manufacturing sectors remains yet relatively low in the EU – 28.3% of people employed in total manufacturing, although this proportion is slightly higher for high tech manufacturing sectors (31.8%).

Figure 6.2. provides the breakdown of the percentage of employment accounted for by medium-high tech manufacturing sectors on the one side, and high tech manufacturing sectors on the other. Of the 7.4% of employment in high tech and medium-high tech manufacturing sectors in the EU, 6.1% corresponded to medium-high tech, whereas 1.3% to high technology. Although Germany remains as the EU Member State most specialised in medium-high tech manufacturing sectors, when looking exclusively at high tech manufacturing sectors, Ireland is ahead, as in 2002 3.2% of its labour force was employed in these sectors. This rate was almost 2.5 times larger than the EU average. The percentage of employment in high tech manufacturing sectors in Germany, France, Austria, Finland, Sweden and the United Kingdom were also equal or above the EU mean.

Among Acceding Countries, the percentage of employment accounted for by high tech manufacturing sectors in Hungary is remarkable: with a rate of 2.6% it outperformed all the EU Member States except for Ireland. The Czech Republic (1.4%) and the Slovak Republic (1.5%) also retained rates above the EU average of 1.3%.

Table 6.2.

Employment in high tech and medium-high tech manufacturing in thousands and percentage of women EU-15 and Candidate Countries 2002 (1)

	Manufacturing		High tech and medium-high tech manufacturing		High tech manufacturing	
	Number of persons employed in thousands	Of which women in %	Number of persons employed in thousands	Of which women in %	Number of persons employed in thousands	Of which women in %
EU-15	31 201	28.3	12 018	24.2	2 126	31.8
BE	742	25.7	271	24.5	40	29.8
DK	444	31.6	173	31.6	30	38.1
DE	8 541	28.2	4 122	23.7	704	32.3
EL	541	29.1	87	20.1	11	: u
ES	3 000	25.7	869	22.3	77	27.2
FR	4 256	29.3	1 628	27.7	314	34.8
IE	283	30.8	120	37.6	55	42.5
IT	4 934	30.4	1 603	22.7	231	32.2
LU	19	18.9	2	34.4 u	1 u	80.2 u
NL	1 031	22.9	332	17.3	89	24.6
AT	729	25.4	246	23.3	67	28.3
PT	1 056	43.9	171	37.6	19	55.9 u
FI	474	29.7	177	25.0	48	35.5
SE	724	26.1	316	25.8	64	36.7
UK	4 428	25.2	1 901	22.5	376	27.4
CZ	1 333	38.3	425	35.5	67	48.9
EE	128	47.9	20	59.2	3	90.7
CY	39	37.1	4	48.4	: u	: u
LV	162	39.9	19	26.0	2	31.2
LT	253	48.9	38	32.5	9	: u
HU	958	40.3	327	35.8	100	49.5
SI	287	39.6	85	38.4	8 u	41.6 u
SK	570	40.8	173	37.0	32	61.1
BG	667	50.9	149	35.1	14	41.0 u
RO	2 101	47.5	537	32.8	34	37.2

Source: Eurostat, EU LFS — spring data.

Figure 6.2.

Percentage of employment in high tech and medium-high tech manufacturing sectors EU-15 and Candidate Countries 2002

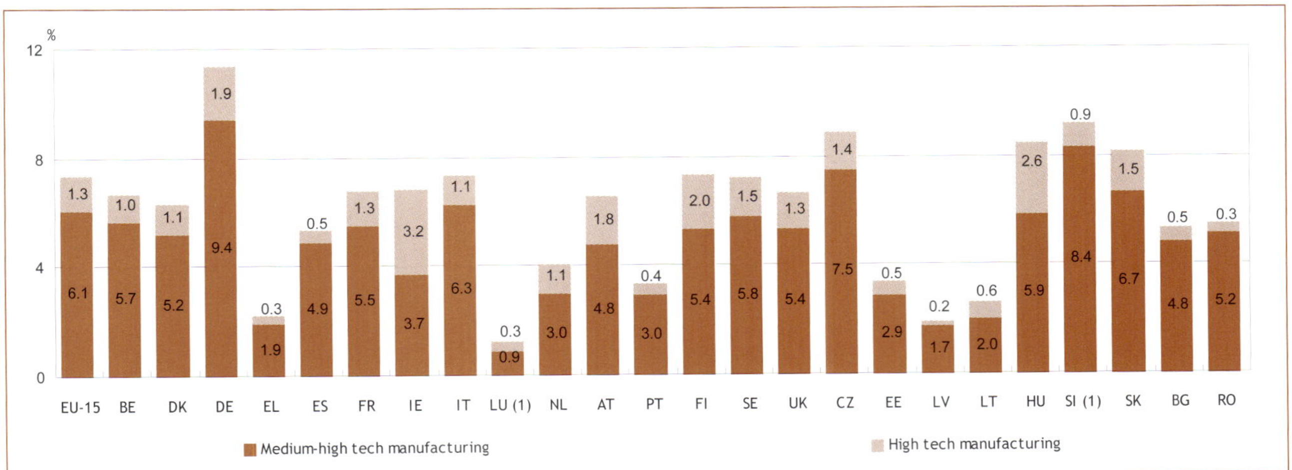

(1) LU and SI — high tech manufacturing: unreliable data.

Source: Eurostat, EU LFS — spring data.

The evolution of employment in high tech and medium-high tech manufacturing sectors is considered in Table 6.3. Employment in high tech and medium-high tech in the EU grew at an annual average growth rate of 0.9% during the 1997-2002 period, compared to 0.4% of overall manufacturing. High tech manufacturing sectors instead grew at an annual average growth rate of 0.3%. Among Member States, Spain (3.9%) shows the highest annual average growth rate for the period 1997-2002, followed by Finland (3.8%); Luxembourg instead showed an average decline of 7.0%.

Hungary, the Acceding Country with the largest high tech manufacturing sector (100 thousand, see Table 6.2.), retained an annual average growth rate of 12.2% in these sectors during the 1997-2002 period. The Czech Republic, with 67 thousand people employed in high tech manufacturing sectors, according to Table 6.2., grew at 4.3% per annum — Table 6.3.

Table 6.3. **Evolution of employment in high tech and medium-high tech manufacturing EU-15 and Candidate Countries 1997 to 2002 (1)**

	Number of persons employed in thousands						Annual average growth rates in %		
	High tech and medium-high tech manufacturing						Manufacturing	High tech and medium-high tech manufacturing	High tech manufacturing
	1997	1998	1999	2000	2001	2002	1997-2002	1997-2002	1997-2002
EU-15	**11 492**	**11 819**	**11 938**	**12 140**	**12 211**	**12 018**	**0.4**	**0.9**	**0.3**
BE	288	279	287	284	265	271	-0.4	-1.2	-2.5
DK	169	183	173	175	190	173	-2.4	0.5	6.6
DE	3 824	3 922	3 924	4 063	4 093	4 122	0.3	1.5	2.3
EL	85	95	87	87	87	87	-0.6	0.5	10.0
ES	717	760	792	825	874	869	3.8	3.9	-0.2 u
FR	1 558	1 551	1 628	1 672	1 695	1 628	0.7	0.9	-0.7
IE	102	112	116	116	125	120	1.6	3.3	6.0
IT	1 449	1 552	1 570	1 596	1 586	1 603	1.9	2.0	1.4
LU	3	3	3	4	2	2	-2.5	-7.0	2.9 u
NL	367	359	355	350	346	332	-0.8	-2.0	0.0
AT	233	235	243	249	240	246	-0.4	1.1	0.5
PT	:	170	173	179	178	171	-2.0	0.2	-0.9
FI	147	157	169	171	179	177	2.3	3.8	4.4
SE	338	340	335	326	335	316	-0.8	-1.3	-3.3
UK	2 052	2 100	2 083	2 043	2 017	1 901	-2.4	-1.5	-3.6
CZ	426	419	415	419	430	425	-0.7	0.0	4.3
EE	28	22	23	24	28	20	-2.7	-6.5	-7.3
CY	:	:	3	3	3	4	1.7	6.4	: u
LV	:	8	9	6	17	19	-4.1	25.4	9.9
LT	:	59	61	48	47	38	-3.4	-10.7	-1.3
HU	282	310	318	307	337	327	2.3	3.0	12.2
SI	79	78	74	78	80	85	0.0	1.4	-3.2 u
SK	:	:	141	143	143	173	1.4	7.2	17.2
BG	:	:	:	161	151	149	-0.4	-3.7	-9.3
RO	751	689	642	543	531	537	-3.0	-6.5	-2.9

(1) **Exceptions to the reference year 1997**
PT, LV and LT: 1998;
CY and SK: 1999;
BG: 2000.

Source: Eurostat, EU LFS — spring data.

Sweden is the Member State most specialised in high tech services — 5.2% of employment — and also in other knowledge-intensive services (41.8%)

In 2002 there were almost 111 million people employed in services in the EU, of which more than 54 million were engaged in knowledge-intensive services — KIS. Among these, almost 6 million people worked in high tech services — Table 6.4.

At the Member State level, the largest number of people employed in knowledge-intensive services in 2002 was retained by the United Kingdom (11 552 thousand), followed by Germany (11 536 thousand), and France (8 485 thousand). Table 6.4. denotes a stronger presence of females in services sectors than in manufacturing — see Table 6.2., especially in knowledge-intensive sectors, where female employment accounts for at least 51.7% of employment not only in the EU Member States but also in each individual Candidate Country. On average, 58.5% of the people employed in KIS in the EU are females. All Candidate Countries except for Cyprus retained rates that were at least 4 percentage points above the EU average.

Table 6.4.

Employment in knowledge-intensive services in thousands and percentage of women EU-15 and Candidate Countries 2002 (1)

	Services		Knowledge-intensive services		High tech services	
	Number of persons employed in thousands	Of which women in %	Number of persons employed in thousands	Of which women in %	Number of persons employed in thousands	Of which women in %
EU-15	**110 737**	**52.2**	**54 257**	**58.5**	**5 803**	**32.5**
BE	2 935	50.7	1 531	57.1	169	25.7
DK	2 011	54.9	1 205	62.5	130	35.2
DE	23 632	55.0	11 536	60.0	1 209	34.0
EL	2 438	43.9	898	51.7	69	27.1
ES	10 189	48.6	4 148	54.9	406	35.0
FR	16 833	53.9	8 485	59.5	971	38.7
IE	1 145	54.3	584	60.1	75	32.4
IT	13 811	45.1	5 973	53.6	657	32.7
LU	147	47.5	72	54.4	4	31.2
NL	6 404	50.1	3 168	57.1	304	25.2
AT	2 444	55.6	1 124	60.6	129	30.9
PT	2 766	54.4	991	65.0	74	37.9
FI	1 622	59.2	944	65.7	114	37.2
SE	3 241	56.8	2 045	63.7	227	34.8
UK	21 120	53.5	11 552	58.3	1 265	27.1
CZ	2 622	54.5	1 138	65.0	147	50.2
EE	364	57.2	179	65.2	17	70.2
CY	226	51.3	83	57.3	6	27.8
LV	584	60.3	244	70.0	21	47.5
LT	767	59.5	351	72.2	24	45.5
HU	2 298	54.2	1 016	64.9	118	44.3
SI	479	54.7	211	62.5	22	33.0
SK	1 167	57.5	507	67.8	60	48.7
BG	1 587	52.9	621	66.2	74	53.3
RO	3 195	50.3	1 254	64.0	153	47.6

Source: Eurostat, EU LFS — spring data.

Figure 6.3. provides the breakdown of the percentage of employment accounted for by high tech services on the one hand, and other knowledge-intensive services on the other. Of the 33.3% KIS employment in the EU, 3.6% corresponded to high tech services and 29.7% to other knowledge-intensive services. The country most specialised in these sectors in 2002 was Sweden, where high tech services accounted for 5.2% of employment and other knowledge-intensive services for 41.8%. Among Acceding Countries, KIS in 2002 still retained ratios of employment below the EU average. Being the Acceding Country with most people employed in KIS – 1 138 thousand, the Czech Republic employed 3.1% and 20.8% of its workforce in high tech services and in other knowledge-intensive services, respectively. However, Estonia was the Acceding Country most specialised in KIS, as these sectors accounted for over 30% of the country's total employment.

Table 6.5. reveals that KIS are the most dynamic sectors in the EU, especially high tech services: for the 1997-2002 period, the EU recorded an annual average growth rate of 3.1% for KIS and 5.6% for high tech services, compared to 2.3% in total services and 0.4% in manufacturing – see Table 6.3. In this period, annual average growth rates for KIS were above their respective growth rates for services for all the Member States. Among Acceding Countries the situation varies: whilst KIS grew faster than services overall in Cyprus, Estonia, Hungary and Slovenia, the Czech Republic and Latvia retained equal rates. On the contrary, employment in KIS in Lithuania and Slovak Republic decreased during the 1997-2002 period.

Figure 6.3.

**Percentage of employment in knowledge-intensive services
EU-15 and Candidate Countries
2002**

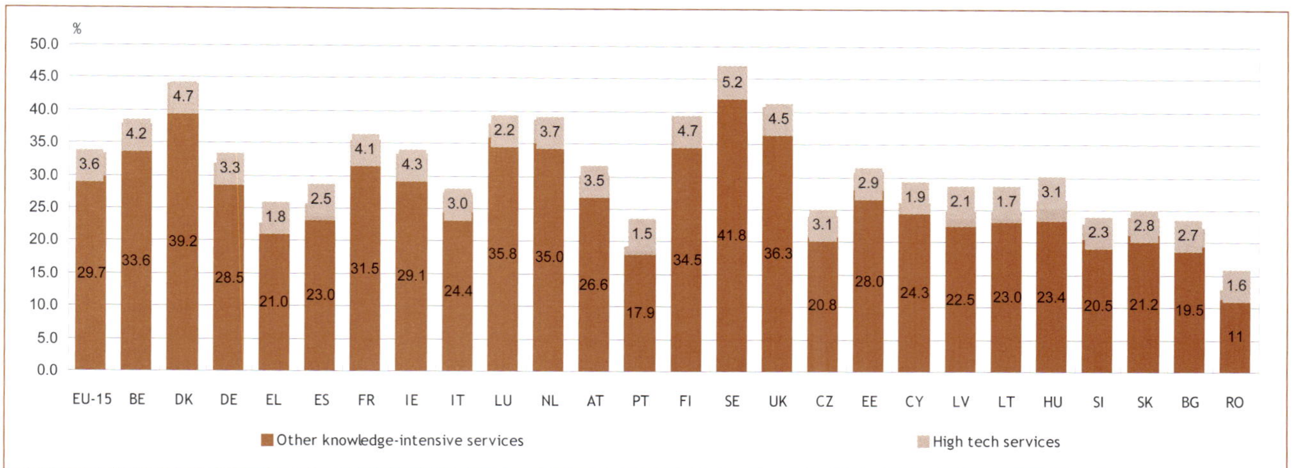

Source: Eurostat, EU LFS — spring data.

Table 6.5.

**Evolution of employment in knowledge-intensive services
EU-15 and Candidate Countries
1997 to 2002 (1)**

	Number of persons employed in thousands						Annual average growth rates in %		
	Knowledge-intensive services						Services	Knowledge-intensive services	High tech services
	1997	1998	1999	2000	2001	2002	1997-2002	1997-2002	1997-2002
EU-15	**46 670**	**48 010**	**49 938**	**51 397**	**53 104**	**54 257**	**2.3**	**3.1**	**5.6**
BE	1 340	1 384	1 464	1 516	1 538	1 531	1.8	2.7	5.7
DK	1 093	1 088	1 125	1 144	1 161	1 205	1.4	2.0	4.4
DE	10 078	10 386	10 797	11 031	11 330	11 536	1.4	2.7	3.9
EL	801	863	872	875	892	898	1.9	2.3	4.7
ES	3 150	3 274	3 483	3 756	3 952	4 148	4.5	5.7	12.3
FR	7 447	7 628	7 814	8 019	8 295	8 485	2.2	2.6	4.4
IE	402	454	497	529	548	584	6.6	7.8	18.2
IT	5 031	5 236	5 404	5 581	5 756	5 973	2.2	3.5	4.6
LU	58	60	67	64	66	72	3.1	4.3	1.4
NL	2 653	2 812	2 970	3 083	3 222	3 168	3.5	3.6	6.8
AT	994	1 016	1 028	1 036	1 082	1 124	1.3	2.5	8.9
PT	:	847	910	925	954	991	3.5	4.0	3.2
FI	792	834	873	898	940	944	3.3	3.6	9.2
SE	1 728	1 732	1 840	1 886	2 002	2 045	3.0	3.4	7.5
UK	10 120	10 395	10 793	11 054	11 365	11 552	2.0	2.7	5.8
CZ	1 121	1 091	1 076	1 124	1 135	1 138	0.3	0.3	-2.4
EE	166	171	166	153	161	179	1.2	1.5	-0.1
CY	:	:	66	70	77	83	5.7	8.0	14.6
LV	:	227	241	240	238	244	1.8	1.8	3.5
LT	:	366	390	400	397	351	-0.9	-1.0	-11.2
HU	908	941	965	1 009	1 006	1 016	1.7	2.3	3.1
SI	180	193	205	203	210	211	2.4	3.2	3.6
SK	:	:	515	510	536	507	0.3	-0.5	0.7
BG	:	:	:	608	637	621	1.0	1.1	1.6
RO	1 313	1 271	1 231	1 181	1 188	1 254	-0.2	-0.9	-3.2

(1) **Exceptions to the reference year 1997**
PT, LV and LT: 1998;
CY and SK: 1999;
BG: 2000.

Source: Eurostat, EU LFS — spring data.

At the regional level

Stuttgart (DE) is the region most specialised in high tech and medium-high tech manufacturing sectors — 21.2% of employment

This section analyses the evolution and composition of employment in high tech and knowledge-intensive sectors in the EU, Iceland and Norway at the regional level. Readers should notice that according to the NUTS classification, for Denmark and Luxembourg the entire national territory is considered as a NUTS 0, 1 and 2 region, which explains their potential appearance in the regional ranking.

Map 6.1. provides an overview of the percentage of employment accounted for by high tech and medium-high tech manufacturing sectors in 2002 across the regions of the EU, Iceland and Norway at the NUTS 2 level. Regions specialised in high tech and medium-high tech manufacturing sectors are highly concentrated in the southern regions of Germany: Baden-Württemberg, Bayern, Rheinhessen-Pfalz and Darmstadt. Braunschweig (DE), Franche-Comté (FR), Alsace (FR) and Piemonte (IT) are also among the leading EU regions in terms of the percentage of employment in high tech and medium-high tech manufacturing sectors.

Figure 6.4. shows the regional disparities in the percentage for employment accounted for by high tech and medium-high tech manufacturing sectors across the European Union. For each Member State, this figure maps the national average, the region with the lowest percentage and the region with the highest percentage.

In 2002 the percentage of employment accounted for by high tech and medium-high tech manufacturing sectors in the EU ranged from 0.8% in Extremadura (ES) to 21.2% in Stuttgart (DE). Belgium, Germany, Spain, France, Italy, Austria, Finland, Sweden and the United Kingdom had at least one region with the percentage of employment accounted for by high tech and medium-high tech manufacturing sectors above the EU average (7.4%). Regional disparities are largest for Germany, France and Italy, with Stuttgart (21.2%), Franche-Comté (17.4%) and Piemonte (13.2%) recording the highest percentage of employment, respectively.

Figure 6.4. **Regional range of percentage of employment accounted for by high tech and medium-high tech manufacturing EU-15 by Member State 2002 (1)**

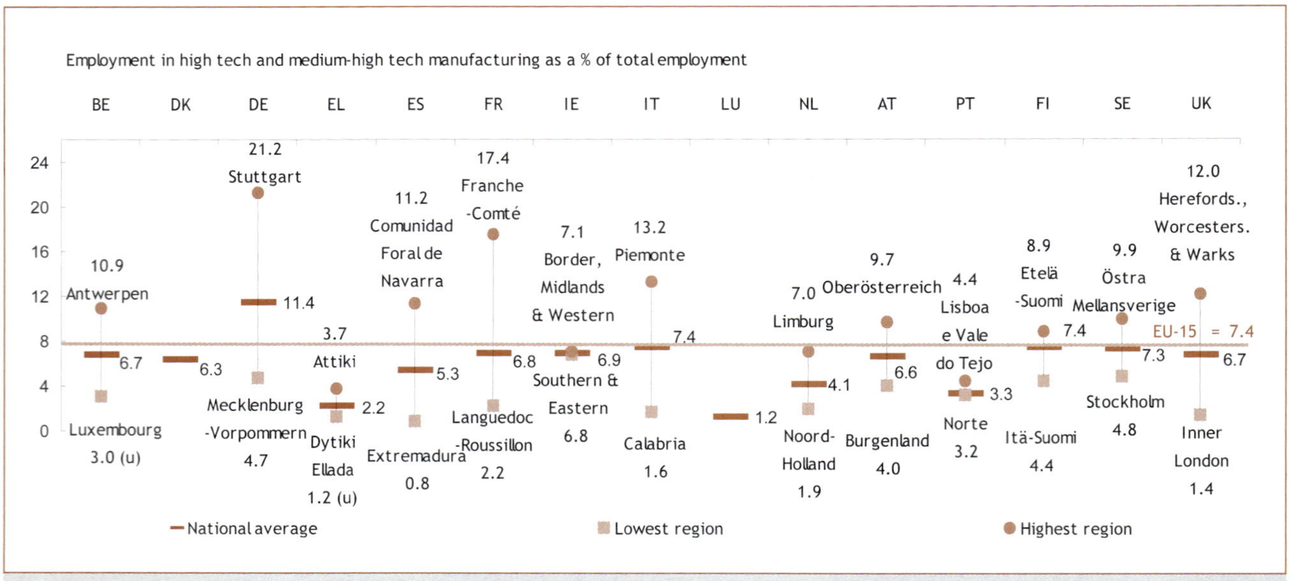

(1) Rankings exclude regions for which reliability levels do not permit publication according to the EU LFS.

Source: Eurostat, EU LFS — spring data.

Map 6.1.

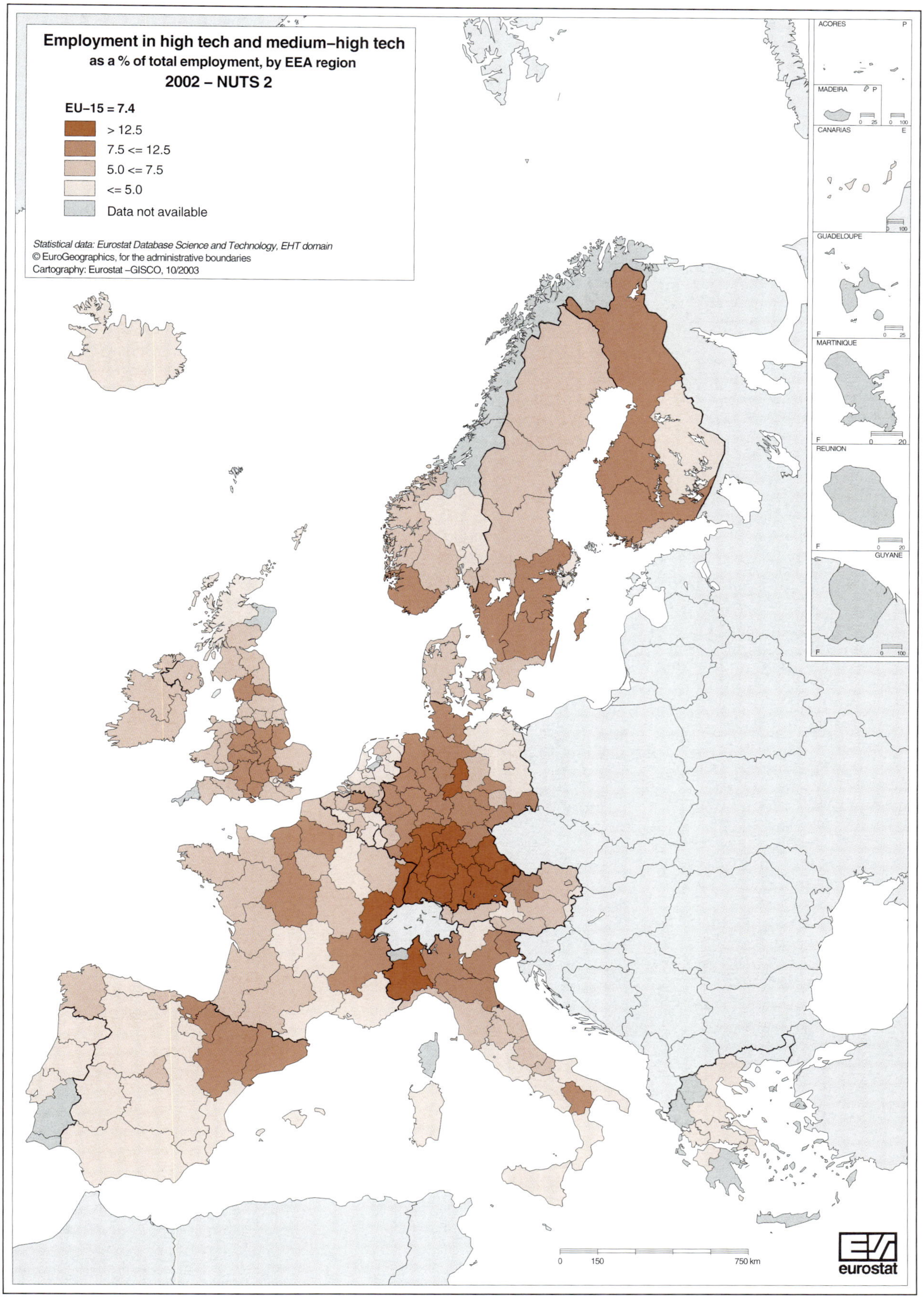

EU-15 = 7.4 refers to the EU-15 average,
i.e. in 2002 7.4% of the EU's workforce was employed in high tech and medium-high tech manufacturing sectors.

Exception to the reference year 2002
IS: 2001.

Luxembourg (BE), Dytiki Ellada (EL), Thessalia (EL), Baleares (ES), Extremadura (ES), La Rioja (ES) and Molise (IT): unreliable data.

Table 6.6. shows the 15 leading NUTS 2 regions in the EU as regards employment in high tech and medium-high tech manufacturing in absolute terms. For these regions, details are provided on employment in total, in manufacturing and in high tech and medium-high tech sectors.

The leading EU region in absolute terms in 2002 was Lombardia (IT), as it employed 431 thousand people in high tech and medium-high tech sectors. These represented 10.7% of employment. During the 1997-2002 period, employment in high tech and medium-high tech manufacturing sectors in Lombardia grew above the EU average (0.9%) at 1.2% per annum. Following Lombardia in the ranking were Stuttgart (DE), the leading EU region in relative terms, with 401 thousand people employed and Cataluña (ES) with 287 thousand. The top 15 regions represented 31% of the EU's total employment in high tech and medium-high tech manufacturing sectors, but only 2% of the total employment in the Union.

Table 6.7. shows the 15 leading NUTS 2 regions in the EU in terms of employment in high tech and medium-high tech manufacturing in relative terms – as a percentage of total employment. As in Table 6.6., details are provided on employment in total, in manufacturing and in high tech and medium-high tech sectors.

The EU region most specialised in high tech and medium-high tech manufacturing sectors in 2002 was Stuttgart (DE), with 21.2% of employment in these sectors. During the 1997-2002 period, employment in high tech and medium-high tech manufacturing sectors in Stuttgart grew at an annual average growth rate of 1.5% above the EU average (0.9%) but below other leading regions such as Tübingen (5.1%). Employment in high tech and in medium-high tech manufacturing sectors in Tübingen accounted for 18.7% of the total employment in the region. The dominance of German regions in high tech and medium-high tech manufacturing as a percentage of total employment is noticeable, as 12 out of the 15 leading regions are situated in this country.

Table 6.6. **Leading EU regions in employment in high tech and medium-high tech manufacturing in absolute terms 2002**

	Country	NUTS 2 region	Total	Manufacturing	High tech and medium-high tech manufacturing			High tech manufacturing		
			In thousands	In thousands	In thousands	% of total employment	AAGR (1) in % 1997-2002 (2)	In thousands	% of total employment	AAGR (1) in % 1997-2002 (2)
		EU-15	162 974	31 201	12 018	7.4	0.9	2 126	1.3	0.3
1	IT	Lombardia	4 011	1 284	431	10.7	1.2	67	1.7	4.2
2	DE	Stuttgart	1 889	658	401	21.2	1.5	63	3.3	-0.8
3	ES	Cataluna	2 769	773	287	10.4	5.2	25	0.9	-1.3
4	FR	Île de France	5 029	564	286	5.7	-2.4	**75**	1.5	-3.6
5	DE	Oberbayern	2 055	486	285	13.9	2.4	54	2.6	5.5
6	DE	Darmstadt	1 753	390	237	13.5	1.5	40	2.3	4.3
7	IT	Piemonte	1 785	533	235	13.2	-0.4	23	1.3	-7.9
8	DE	Düsseldorf	2 200	505	209	9.5	-1.5	34	1.6	0.0
9	FR	Rhône-Alpes	2 376	511	205	8.6	1.5	48	2.0	1.0
10	DE	Karlsruhe	1 245	359	204	16.4	-0.1	36	2.9	-1.6
11	IT	Veneto	1 972	631	197	10.0	2.8	33	1.7	8.8
12	DE	Köln	1 832	400	197	10.7	0.6	32	1.7	-0.5
13	IT	Emilia-Romagna	1 804	499	188	10.4	2.6	20	1.1	0.7
14	DK	Denmark	2 741	444	173	6.3	0.5	30	1.1	6.6
15	DE	Arnsberg	1 559	432	163	10.4	1.9	27	1.7	7.8

(1) AAGR — Annual average growth rate.
(2) **Exceptions to the reference period 1997-2002**
Chemnitz, Dresden and Leipzig (all in DE): 2000-2002;
PT regions: 1998-2002.

Source: Eurostat, EU LFS — spring data.

Table 6.7. **Leading EU regions in employment in high tech and medium-high tech manufacturing as a percentage of total employment (1) 2002**

	Country	NUTS 2 region	Total	Manufacturing	High tech and medium-high tech manufacturing			High tech manufacturing		
			In thousands	In thousands	In thousands	% of total employment	AAGR (2) in % 1997-2002 (3)	In thousands	% of total employment	AAGR (2) in % 1997-2002 (3)
		EU-15	162 974	31 201	12 018	7.4	0.9	2 126	1.3	0.3
1	DE	Stuttgart	1 889	658	401	21.2	1.5	63	3.3	-0.8
2	DE	Tübingen	845	287	158	18.7	5.1	36	4.3	11.6
3	DE	Braunschweig	687	198	121	17.5	3.2	12	1.8	10.4
4	FR	Franche-Comté	503	161	88	17.4	2.9	19	3.8	2.1
5	DE	Karlsruhe	1 245	359	204	16.4	-0.1	36	2.9	-1.6
6	DE	Niederbayern	574	172	90	15.6	3.8	10	1.7	-1.9
7	DE	Unterfranken	619	191	96	15.6	1.9	14	2.3	4.4
8	DE	Rheinhessen-Pfalz	897	230	138	15.4	-0.6	12	1.4	1.1
9	DE	Freiburg	1 010	323	151	14.9	4.7	49	4.9	6.2
10	DE	Schwaben	845	248	122	14.5	3.3	18	2.1	8.2
11	DE	Mittelfranken	786	214	111	14.1	-0.7	20	2.5	-3.4
12	DE	Oberbayern	2 055	486	285	13.9	2.4	54	2.6	5.5
13	DE	Darmstadt	1 753	390	237	13.5	1.5	40	2.3	4.3
14	IT	Piemonte	1 785	533	235	13.2	-0.4	23	1.3	-7.9
15	FR	Alsace	767	205	100	13.0	2.7	11	1.5	0.0

(1) NUTS 2 regions are only taken into account if there are at least 80 thousand people working in high tech and medium-high tech manufacturing sectors.
(2) AAGR — Annual average growth rate.
(3) **Exceptions to the reference period 1997-2002**
Chemnitz, Dresden and Leipzig (all in DE): 2000-2002;
PT regions: 1998-2002.

Source: Eurostat, EU LFS — spring data.

During the 1997-2002 period, employment in high tech and medium-high tech manufacturing in the EU grew at an annual average growth rate of 0.9%. Table 6.8. shows the regions with the highest growth and the regions with the lowest growth in these sectors. The reader should note that, in order to avoid biases in the data presented, only regions with at least 80 thousand people working in high tech industries are taken into account for the ranking. According to Table 6.8., the most dynamic EU region during the 1997-2002 period was Southern and Eastern (IE), as it grew at an annual average growth rate of 17.0%. Following Southern and Eastern were Thüringen (DE) with a rate of 6.2% and Cataluña (ES) growing at 5.2%.

The regions where employment in high tech and medium-high tech manufacturing decreased the most during the 1997-2002 period were Gloucestershire, Wiltshire and North Sommerset in the United Kingdom (-3.6%) and Schleswig-Holstein in Germany (-3.4%).

Figures 6.5. and 6.6. show the leading regions in high technology manufacturing sectors alone.

In absolute terms, Île de France (FR) was the region that employed most people, 75 thousand, followed by Lombardia (IT) with 67 thousand and Stuttgart (DE) with 63 thousand – Figure 6.5. The leading ten regions in employment in high tech manufacturing sectors accounted for 24% of the EU's total employment in these sectors, but it barely represented a 0.4% of EU-15's total employment.

As a percentage of total employment, Freiburg (DE) was leading, as 4.9% of the people employed in this region were working in high tech manufacturing sectors compared to 1.3% in the EU overall. Following Freiburg were Tübingen (DE) and Franche-Comté (FR), with 4.3% and 3.8% of employment accounted for by high tech sectors, respectively – Figure 6.6.

Table 6.8.

EU regions with highest and lowest growth (1) in employment in high tech and medium-high tech manufacturing 1997 to 2002

Regions with highest growth					Regions with lowest growth				
Country	NUTS 2 region	In thousands 2002	% of total employment 2002	AAGR (2) in % 1997-2002 (3)	Country	NUTS 2 region	In thousands 2002	% of total employment 2002	AAGR (2) in % 1997-2002 (3)
IE	Southern and Eastern	89	6.8	17.0	UK	Gloucesters., Wilts. & North Somerset	86	7.6	-3.6
DE	Thüringen	94	8.9	6.2	DE	Schleswig-Holstein	93	7.6	-3.4
ES	Cataluna	287	10.4	5.2	DE	Hannover	87	9.5	-3.0
DE	Tübingen	158	18.7	5.1	UK	West Midlands	120	10.6	-2.7
DE	Weser-Ems	96	9.2	4.8	FR	Île de France	286	5.7	-2.4
DE	Freiburg	151	14.9	4.7	DE	Berlin	85	5.9	-2.4
DE	Münster	106	9.9	4.0	UK	Greater Manchester	86	7.2	-1.9
DE	Niederbayern	90	15.6	3.8	NL	Noord-Brabant	84	6.8	-1.8
FR	Bretagne	80	6.4	3.8	SE	Västsverige	81	9.3	-1.7
DE	Schwaben	122	14.5	3.3	DE	Düsseldorf	209	9.5	-1.5

NB: Employment in high tech and medium-high tech manufacturing in the EU grew during the 1997-2002 period at an annual average growth rate of 0.9%.
(1) NUTS 2 regions are only taken into account if there are at least 80 thousand people working in high tech and medium-high tech manufacturing sectors.
(2) AAGR — Annual average growth rate.
(3) **Exceptions to the reference period 1997-2002**
Chemnitz, Dresden and Leipzig (all in DE): 2000-2002;
PT regions: 1998-2002.

Source: Eurostat, EU LFS — spring data.

Figure 6.5.

Leading EU regions in employment in high tech manufacturing in absolute terms 2002

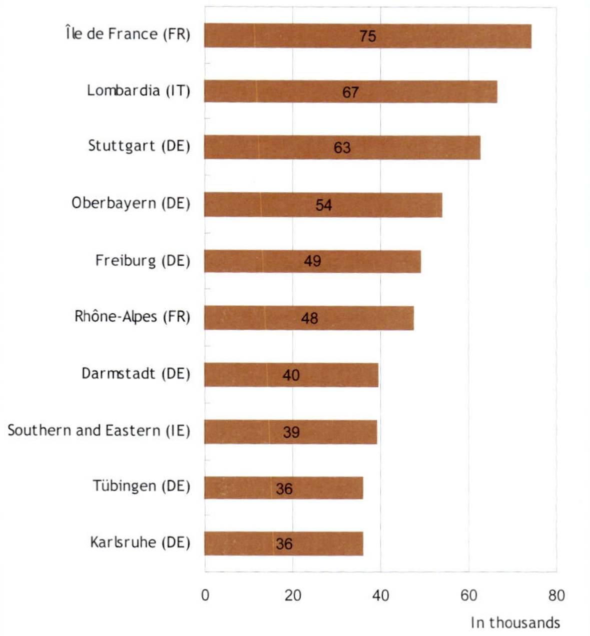

Source: Eurostat, EU LFS — spring data.

Figure 6.6.

Leading EU regions in employment in high tech manufacturing as a percentage of total employment (1) 2002

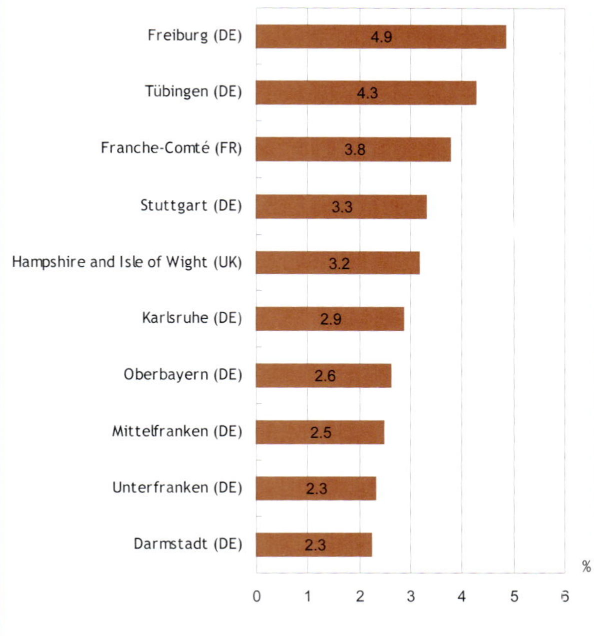

(1) NUTS 2 regions are only taken into account if there are at least 80 thousand people working in high tech and medium-high tech manufacturing sectors.

Source: Eurostat, EU LFS — spring data.

Inner London (UK) is the EU region most specialised in employment in knowledge-intensive services — 59.1% of employment

Map 6.2. shows employment in knowledge-intensive services — KIS — as a percentage of total employment for the regions of the EU, Iceland and Norway at the NUTS 2 level. It can be seen that regions where employment is most specialised in KIS are more evenly distributed across Europe than the leading regions in employment in high tech and medium-high tech showed in Map 6.1. However, it may also be observed that most specialised regions tend to be concentrated around the main European cities such as London, Paris, Brussels, Amsterdam, Copenhagen, Berlin, Hamburg and Wien. In the case of Norway and Sweden, it is not only the region of the capital city where KIS account for a large proportion of employment, but almost the entire country is highly specialised in these sectors. This may be due to the role of the public services sector, which tends to be very high in these countries, especially in social and health services sectors.

Figure 6.7. shows the regional disparities in the percentage for employment accounted for by knowledge-intensive services across the EU. For each Member State, this figure maps the national average, the region with the lowest percentage and the region with the highest percentage.

The proportion of employment accounted for by KIS sectors in the EU ranged from 11.1% in Sterea Ellada (EL) to 59.1% in Inner London (UK). With the exception of Greece and Portugal, all countries had at least one region with the percentage of employment accounted for by knowledge-intensive services above the EU average (33.3%).

Figure 6.7. **Regional range of percentage of employment accounted for by knowledge-intensive services — KIS EU-15 by Member State 2002 (1)**

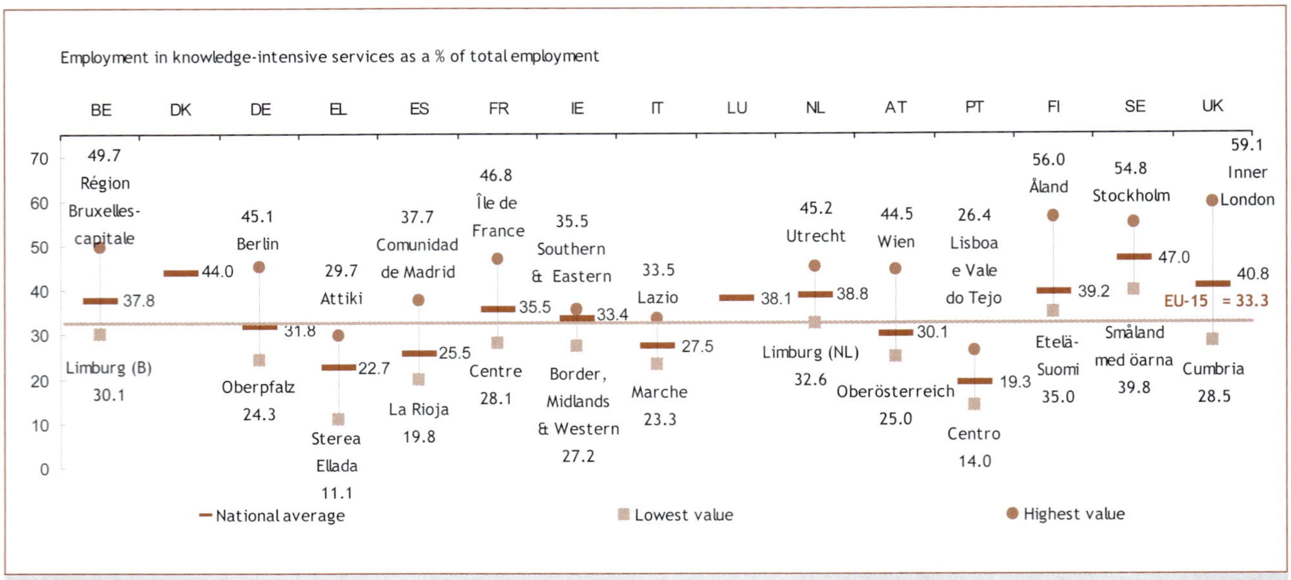

(1) Rankings exclude regions for which reliability levels do not permit publication according to the EU LFS.

Source: Eurostat, EU LFS — spring data.

Map 6.2.

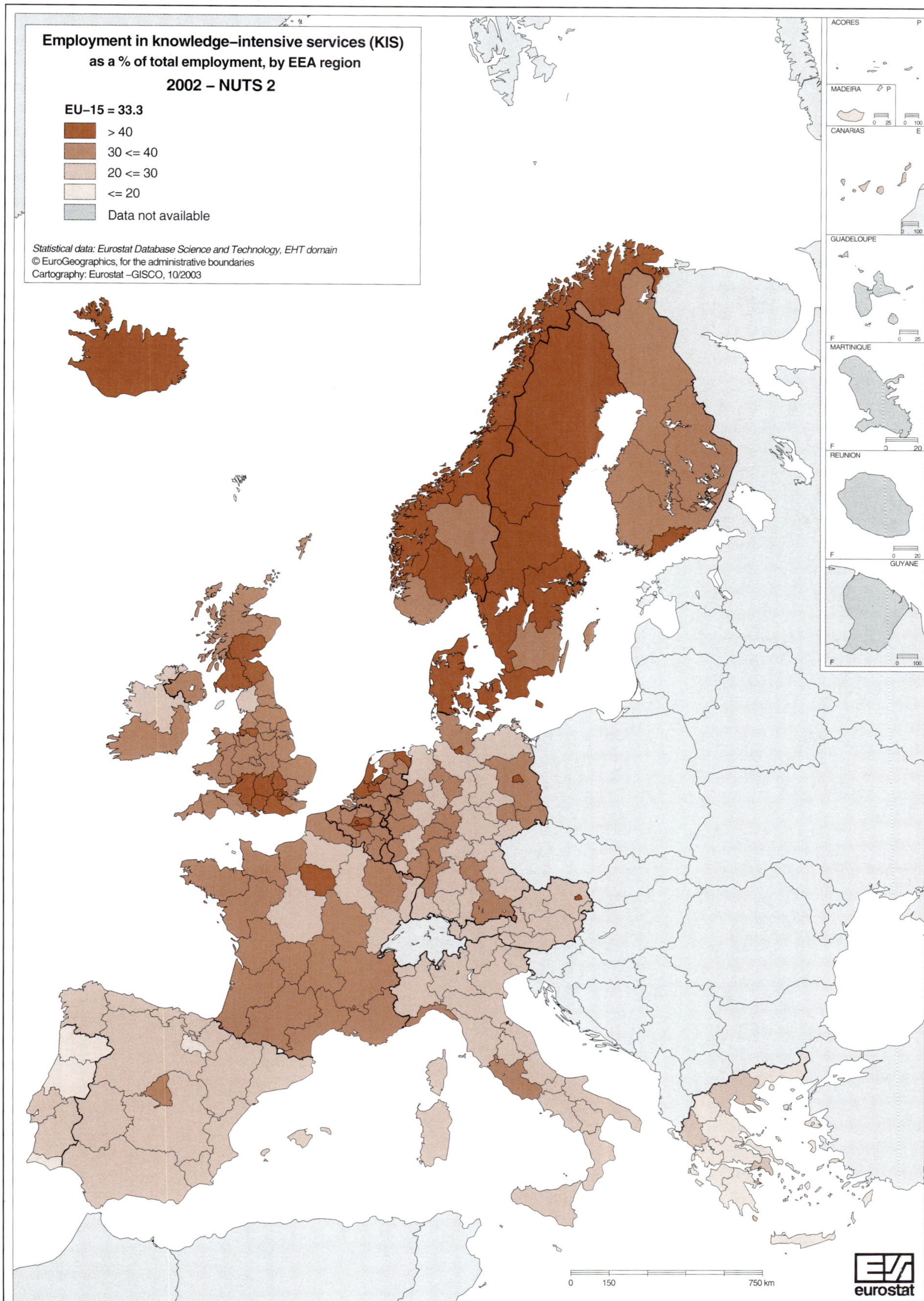

EU-15 = 33.3 refers to the EU-15 average,
i.e. in 2002 33.3% of the EU's workforce was employed in knowledge-intensive services.

Exception to the reference year 2002
IS: 2001.

Table 6.9. shows the 15 leading NUTS 2 regions in the EU as regards employment in knowledge-intensive services in absolute terms. For these regions, details are provided on employment in total, in services, in knowledge-intensive services and in high tech services.

The leading EU region in absolute terms in 2002 was Île de France (FR), as it employed 2 353 thousand people in knowledge-intensive services. These represented 46.8% of the region's employment. During the 1997-2002 period, employment in knowledge-intensive services in Île de France grew above the EU average (3.1%) at 3.4% per annum. Following Île de France in the ranking were Denmark (DK) with 1 205 thousand people employed and Lombardia (IT) with 1 119 thousand. The top 15 regions amounted to 25% of the EU's total employment in knowledge-intensive services, which represented 8% of the Union's total workforce.

Table 6.10. shows the 15 leading NUTS 2 regions in the EU in terms of employment in knowledge-intensive services as a percentage of total employment. As in Table 6.9., details are provided on employment in total, in services, in knowledge-intensive services and in high tech services.

The EU region most specialised in knowledge-intensive services in 2002 was Inner London (UK), with 59.1% of employment in these sectors. During the 1997-2002 period, employment in knowledge-intensive services in Inner London grew above the EU average (3.1%) at an annual average growth rate of 4.3%. Following Inner London in this ranking were Stockholm (54.8% of employment) and Outer London (50.3%).

Table 6.9. **Leading EU regions in employment in knowledge-intensive services — KIS in absolute terms 2002**

	Country	NUTS 2 region	Total	Services	Knowledge-intensive services			High tech services		
			In thousands	In thousands	In thousands	% of total employment	AAGR (1) in % 1997-2002 (2)	In thousands	% of total employment	AAGR (1) in % 1997-2002 (2)
	EU-15		162 974	110 737	54 257	33.3	3.1	4 418	3.6	5.6
1	FR	Île de France	5 029	4 144	2 353	46.8	3.4	393	7.8	8.0
2	DK	Denmark	2 741	2 011	1 205	44.0	2.0	130	4.7	4.4
3	IT	Lombardia	4 011	2 304	1 119	27.9	3.4	144	3.6	4.5
4	UK	Outer London	2 221	1 876	1 118	50.3	2.3	153	6.9	6.8
5	ES	Comunidad de Madrid	2 318	1 728	874	37.7	8.4	153	6.6	19.1
6	FR	Rhône-Alpes	2 376	1 600	827	34.8	1.2	93	3.9	1.7
7	UK	Inner London	1 332	1 187	788	59.1	4.3	68	5.1	5.5
8	DE	Oberbayern	2 055	1 363	736	35.8	2.6	109	5.3	5.8
9	ES	Cataluna	2 769	1 629	716	25.8	4.6	77	2.8	10.8
10	NL	Zuid-Holland	1 725	1 416	700	40.6	3.3	80	4.6	6.7
11	DE	Düsseldorf	2 200	1 479	691	31.4	2.7	81	3.7	2.8
12	IT	Lazio	2 039	1 585	684	33.5	3.5	116	5.7	5.5
13	DE	Darmstadt	1 753	1 223	673	38.4	3.5	90	5.1	5.4
14	DE	Berlin	1 448	1 159	652	45.1	1.8	75	5.2	5.4
15	DE	Köln	1 832	1 279	640	34.9	2.7	73	4.0	2.8

(1) AAGR — Annual average growth rate.
(2) **Exceptions to the reference period 1997-2002**
Chemnitz, Dresden and Leipzig (all in DE): 2000-2002;
PT regions: 1998-2002.

Source: Eurostat, EU LFS — spring data.

Table 6.10. **Leading EU regions in employment in knowledge-intensive services — KIS as a percentage of total employment (1) 2002**

	Country	NUTS 2 region	Total	Services	Knowledge-intensive services			High tech services		
			In thousands	In thousands	In thousands	% of total employment	AAGR (2) in % 1997-2002 (3)	In thousands	% of total employment	AAGR (2) in % 1997-2002 (3)
	EU-15		162 974	110 737	54 257	33.3	3.1	4 418	3.6	5.6
1	UK	Inner London	1 332	1 187	788	59.1	4.3	68	5.1	5.5
2	SE	Stockholm	969	831	532	54.8	4.1	85	8.8	8.6
3	UK	Outer London	2 221	1 876	1 118	50.3	2.3	153	6.9	6.8
4	FR	Île de France	5 029	4 144	2 353	46.8	3.4	393	7.8	8.0
5	FI	Uusimaa (Suuralue)	749	592	349	46.5	4.6	59	7.9	13.0
6	UK	Surrey, East & West Sussex	1 308	1 037	602	46.0	1.6	77	5.9	3.6
7	SE	Västsverige	870	632	395	45.5	6.1	40	4.6	8.5
8	DE	Berlin	1 448	1 159	652	45.1	1.8	75	5.2	5.4
9	SE	Östra Mellansverige	716	509	318	44.5	2.9	36	5.0	4.6
10	AT	Wien	746	595	332	44.5	3.1	52	7.0	15.2
11	NL	Noord-Holland	1 345	1 132	593	44.1	3.7	52	3.9	6.3
12	DK	Denmark	2 741	2 011	1 205	44.0	2.0	130	4.7	4.4
13	UK	Berks., Bucks & Oxfords.	1 165	884	510	43.8	2.5	99	8.5	6.1
14	DE	Hamburg	792	622	344	43.4	2.6	34	4.3	7.0
15	UK	Eastern Scotland	914	689	388	42.5	3.6	28	3.1	-1.8

(1) NUTS 2 regions are only taken into account if there are at least 300 thousand people working in KIS.
(2) AAGR — Annual average growth rate.
(3) **Exceptions to the reference period 1997-2002**
Chemnitz, Dresden and Leipzig (all in DE): 2000-2002;
PT regions: 1998-2002.

Source: Eurostat, EU LFS — spring data.

During the 1997-2002 period, employment in knowledge-intensive services in the EU grew at an annual average growth rate of 3.1%. Table 6.11. shows the regions with the highest growth and the regions with the lowest growth in these sectors. The reader should note that, in order to avoid biases in the data presented, only regions with at least 300 thousand people working in knowledge-intensive services are taken into account for the ranking. According to Table 6.11., the most dynamic EU region during the 1997-2002 period was Southern and Eastern (IE), as it grew at an annual average growth rate of 31.5%. Following Southern and Eastern were Comunidad de Madrid (ES) with a rate of 8.4% and Lorraine (FR) growing at 6.4%.

The regions where employment in knowledge-intensive services grew the least during the 1997-2002 period were Kent (UK), Rhônes-Alpes (FR) and Bedfordshire, Hertfordshire (UK), which even if below the EU average, still showed a positive trend (0.9% and above).

Table 6.11. **EU regions with highest and lowest growth (1) in employment in knowledge-intensive services — KIS 1997 to 2002**

Regions with highest growth					Regions with lowest growth				
Country	NUTS 2 region	In thousands 2002	% of total employment 2002	AAGR (2) in % 1997-2002 (3)	Country	NUTS 2 region	In thousands 2002	% of total employment 2002	AAGR (2) in % 1997-2002 (3)
IE	Southern and Eastern	465	35.5	31.5	UK	Kent	304	39.8	0.9
ES	Comunidad de Madrid	874	37.7	8.4	FR	Rhône-Alpes	827	34.8	1.2
FR	Lorraine	347	33.7	6.4	UK	Bedfordshire, Hertfordshire	346	41.4	1.3
SE	Västsverige	395	45.5	6.1	UK	West Yorkshire	396	40.0	1.4
ES	Comunidad Valenciana	395	22.7	6.0	UK	South Western Scotland	390	40.5	1.5
ES	Andalucia	593	23.5	5.6	UK	Surrey, East and West Sussex	602	46.0	1.6
DE	Brandenburg	340	30.5	5.1	DE	Hannover	307	33.6	1.6
UK	Hamps. & Isle of Wight	381	41.3	4.8	UK	Essex	324	39.9	1.7
IT	Emilia-Romagna	478	26.5	4.7	FR	Provence-Alpes-Côte d'Azur	571	35.9	1.7
ES	Cataluna	716	25.8	4.6	DE	Berlin	652	45.1	1.8

NB: Employment in knowledge-intensive services in the EU grew during the 1997-2002 period at an annual average growth rate of 3.1%.

(1) NUTS 2 regions are only taken into account if there are at least 300 thousand people working in KIS.

(2) AAGR — Annual average growth rate.

(3) **Exceptions to the reference period 1997-2002**
Chemnitz, Dresden and Leipzig (all in DE): 2000-2002;
PT regions: 1998-2002.

Source: Eurostat, EU LFS — spring data.

Figures 6.8. and 6.9. show the leading regions when high technology services are only taken into account.

In absolute terms, Île de France (FR) was the region that employed most people, 393 thousand, followed by Comunidad de Madrid (ES) and Outer London (UK) both with 153 thousand — Figure 6.8. The leading ten regions in employment in high tech services sectors accounted for 33% of the EU's total.

As a percentage of total employment, Stockholm (SE) was leading, as 8.8% of the people employed in this region were working in high tech services sectors compared to the EU average of 3.6%. Following Stockholm were Berkshire, Bucks and Oxfordshire (UK) and Uusimaa (Suuralue) in Finland, with 8.5% and 7.9% of employment in high tech services, respectively — Figure 6.9.

Figure 6.8. **Leading EU regions in employment in high tech services in absolute terms 2002**

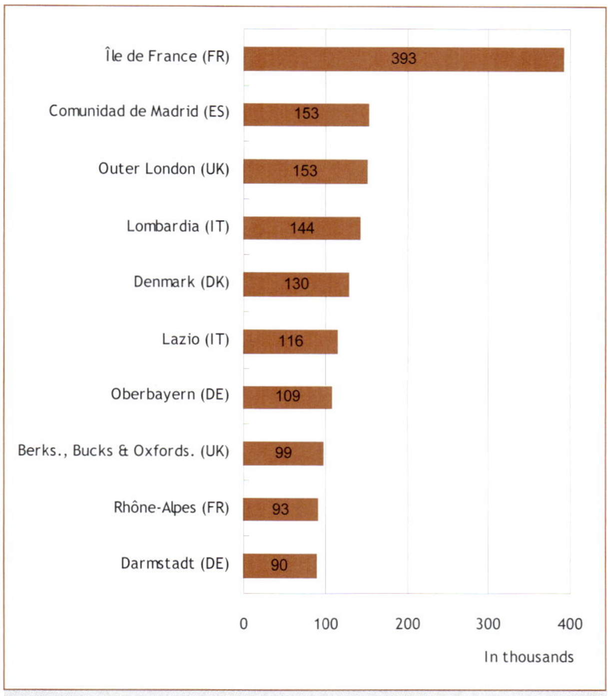

Source: Eurostat, EU LFS — spring data.

Figure 6.9. **Leading EU regions in employment in high tech services as a percentage of total employment (1) 2002**

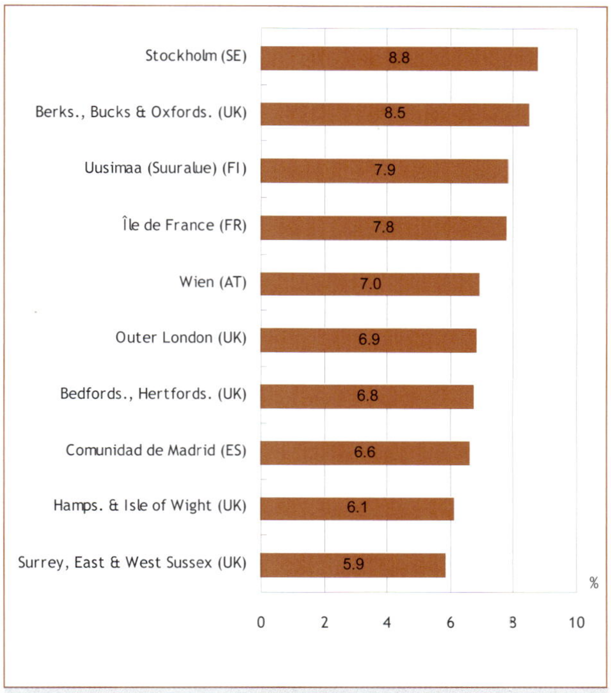

(1) NUTS 2 regions are only taken into account if there are at least 300 thousand people working in KIS.

Source: Eurostat, EU LFS — spring data.

6.3. Value added and labour productivity of high tech sectors

Table 6.12. shows the value added generated and the corresponding labour productivity in selected sectors in the EU and Candidate Countries. In 2000, manufacturing sectors recorded a value added of almost EUR 1 500 thousand million, of which EUR 453 thousand million corresponded to medium-high tech manufacturing sectors and EUR 201 thousand million to high tech manufacturing sectors. Whilst value added in knowledge-intensive market services amounted to EUR 848 thousand million, that of high tech services reached almost EUR 331 thousand million.

Whilst the labour productivity rate for overall manufacturing in the EU was EUR 52 thousand per person employed, high tech manufacturing sectors registered a rate of EUR 73 thousand per person employed. High tech services had a labour productivity of EUR 68 thousand per person employed and medium-high tech manufacturing sectors EUR 58 thousand. Knowledge-intensive market services, in turn, retained a labour productivity rate of EUR 53 thousand per person employed.

At the Member State level, Ireland recorded the highest labour productivity in manufacturing – EUR 132 thousand per person employed, high tech manufacturing – EUR 154 thousand per person employed – and medium-high tech manufacturing – EUR 244 thousand per person employed. Germany, Luxembourg and Denmark led in knowledge-intensive market services: EUR 100, EUR 60 and EUR 59 thousand per person employed, respectively. Luxembourg, instead, was the most productive Member State in high tech services – EUR 124 thousand per person employed.

Regarding Candidate Countries, productivity levels are still below the EU average, with the exception of Malta which registered relatively high rates for high tech manufacturing – EUR 73 thousand per person employed, knowledge-intensive market services – EUR 76 thousand per person employed – and high tech services – EUR 50 thousand per person employed. Malta was also the Acceding Country with the highest labour productivity in medium-high tech manufacturing sectors – EUR 33 thousand per person employed.

Table 6.12.

Value added and labour productivity in selected sector
EU-15 and Candidate Countries
2000

	Manufacturing		Medium-high tech manufacturing		High tech manufacturing		Knowledge-intensive market services		High tech services	
	Value added at factor cost	Labour productivity	Value added at factor cost	Labour productivity	Value added at factor cost	Labour productivity	Value added at factor cost	Labour productivity	Value added at factor cost	Labour productivity
	Mio EUR	Thousand EUR per person employed	Mio EUR	Thousand EUR per person employed	Mio EUR	Thousand EUR per person employed	Mio EUR	Thousand EUR per person employed	Mio EUR	Thousand EUR per person employed
EU-15 (1)	**1 468 402**	**52**	**453 370**	**58**	**201 121**	**73**	**848 176 u**	**53 u**	**330 534 u**	**68 u**
BE	44 482	66	13 884	78	5 534	103	17 081	41	8 938	68
DK	24 599	50	5 903	51	3 511	79	15 745	59	5 995	56
DE	405 409	54	172 378	60	46 320	65	283 147 u	100 u	88 356 u	107 u
EL (2)	8 901	40	1 280	40	552	44	:	:	:	:
ES	100 442	39	26 230	47	6 666	57	51 091	30	17 275	50
FR	210 339	52	57 891	59	37 521	72	113 613	47	45 988	57
IE	33 812	132	11 659	244	10 201	154	5 617	48	5 600	120
IT	204 184	42	55 417	48	19 988	58	59 139	34	34 683	57
LU	2 339	68	344	69	72	44	1 698	60 u	1 264	124
NL	56 861	62	14 961	72	6 762	69	44 872	38	16 174	52
AT	35 558	57	9 750	63	4 018	71	12 508	51	5 401	53
PT	18 127	19	3 378	26	1 134	35	6 594	24	3 516	65
FI	30 748	71	5 745	59	7 301	126	7 757	46	4 135	51
SE	48 951	62	16 171	65	7 816	76	23 421 u	51 u	11 352 u	55 u
UK	243 650	59	58 379	60	43 725	82	205 893	56	81 857	69
CZ	13 391	11	4 360	12	877	12	2 580	11	2 206	21
EE	854	7	106	9	64	7	347	9	241	20
CY	975	29	76	26	35	42	:	:	:	:
LV	1 068	7	90	6	36	8	501	11	382	18
LT	:	:	:	:	:	:	338	7	326	15
HU (2)	9 342	12	3 146	16	1 333	17	1 183	10	1 957	23
MT (3)	930	32	76	33	353	73	327	76	175	50
PL	33 107	14	6 926	14	1 882	17	9 993	23	6 149	23
SI (4)	3 532	15	881	16	475	23	640	16	233	22
SK	3 059	7	856	7	173	7	574	9	595	12
BG	1 819	3	384	3	121	5	225	3	528	10
RO	6 433	4	1 421	4	319	7	656	5	1 408	10

NB: Cells flagged as 'u' refer to values partly estimated, and hence, their quality might be inferior.

(1) Knowledge-intensive market services and high tech services: EU-15 excludes EL.

(2) Exceptions to the minimum enterprise size — number of persons employed
EL: 10; HU: 5.

(3) MT: high technology services excludes K73, as no data are available for these sectors.

(4) SI: high technology services excludes K72 and K73, as no data are available for these sectors.

Source: Eurostat, SBS.

6.4. International trade of high technology products

Global high tech trends

In 2001, high tech exports accounted for 29%, 25% and 20% of total exports in the United States, Japan and the EU, respectively. High tech imports accounted for 19% of total imports for the United States and Japan and 21% for the EU

The evolution of the proportion of trade represented by high tech products between 1996 and 2001 in the EU, Acceding Countries, Japan and the United States is shown in Figures 6.10. (exports) and 6.11. (imports). During this period, high tech exports for the United States accounted for 26-30% of their total exports. This proportion was between 25-27% for Japan, but lower for EU-15 – between 16-20% – and even less for the Acceding Countries – between 5-11%.

The proportion of high tech exports accounted for by high tech products during the 1996-2001 period increased for the EU, Acceding Countries and the United States. However, a slight decrease was registered by Japan.

Looking at imports in Figure 6.11., less than 21% of the total imports for the United States and Japan were accounted for by high tech products during the 1996-2001 period. For the EU, this was also the case up to 1997 after which the high tech proportion of total imports rose up to nearly 23% in 2000 and declined in 2001 to 21%. High tech imports for the Acceding Countries accounted between 10% and 14% of total imports.

Figure 6.12. shows the high tech exports and imports for the EU and Acceding Countries aggregates as well as the world leading countries in 2001. The United States appears as the leading country in the world both as an exporter and as an importer of high tech products closely followed by the EU. However, amounting to EUR 23 thousand million, the EU had the largest high tech trade deficit.

Japan is the third leading high tech exporter, but it has the highest high tech trade balance surplus of EUR 39 thousand million. Among the EU countries, Germany, France, United Kingdom and the Netherlands feature in the top exporters and importers of high tech products, all with a positive high tech trade balance.

In the United States and Japan, high tech imports as a percentage of total imports are for every year below the corresponding percentage of high tech exports – see Figures 6.10. and 6.11., resulting thus in a positive trade balance for Japan and a minor negative balance for the United States. However, the reverse is true for the EU Member States and the Acceding Countries, where the high tech imports percentage is always slightly higher than the corresponding exports percentage, leading thus to negative high tech trade balances as shown in Figure 6.12.

Figure 6.10.

**High tech exports as a percentage of total exports
EU-15, Acceding Countries, Japan and the United States
1996 to 2001**

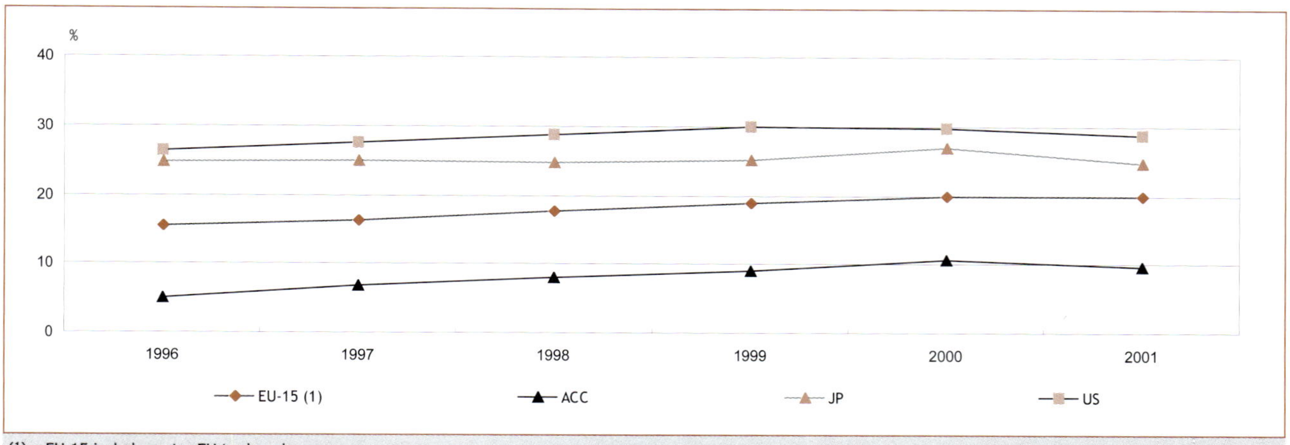

(1) EU-15 includes extra-EU trade only.

Sources: Eurostat, Comext; UN, Comtrade.

Figure 6.11.

**High tech imports as a percentage of total imports
EU-15, Acceding Countries, Japan and the United States
1996 to 2001**

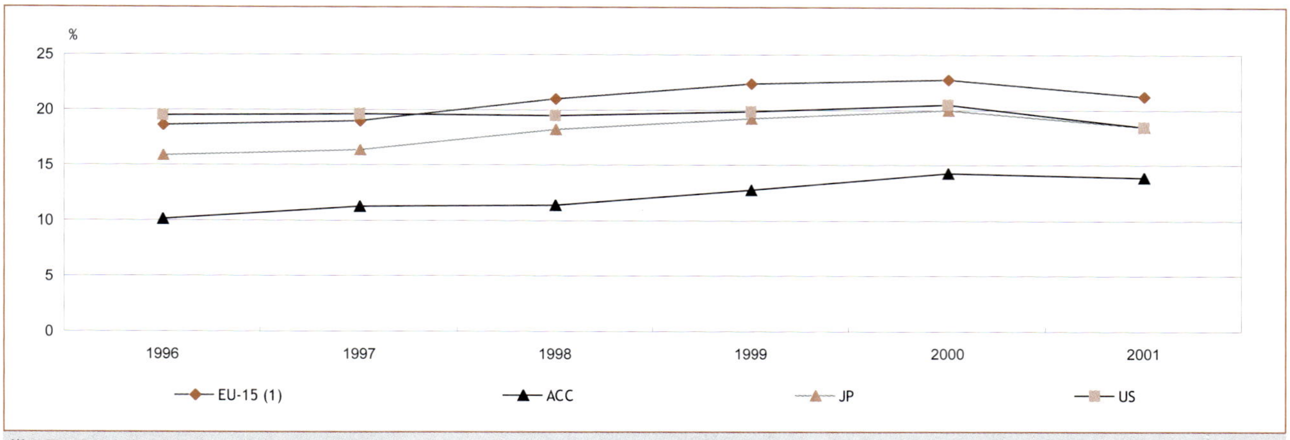

(1) EU-15 includes extra-EU trade only.

Sources: Eurostat, Comext; UN, Comtrade.

Figure 6.12.

**World leading countries
in high tech exports and imports
2001**

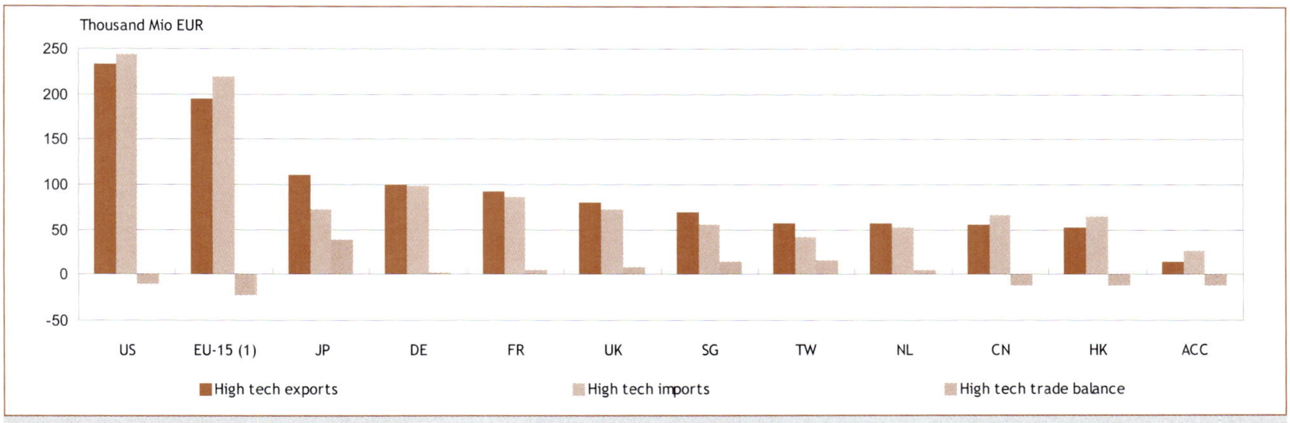

(1) EU-15 includes extra-EU trade only.

Sources: Eurostat, Comext; UN, Comtrade.

Figure 6.13. examines the annual average growth rates – AAGR – of high tech exports and imports during the 1996-2001 period. The EU experienced almost the same growth for both exports – 15.0% per annum – and imports – 15.1% per annum. This rate was above those retained by Japan and the United States during the same period, where imports grew faster than exports. Whilst the annual average growth rate recorded by Japan for high tech imports was equal to 10.5%, exports grew at 6.7% per annum. United States high tech imports and exports grew at 14.1% and 12.6%, respectively. The Acceding Countries recorded a high annual average growth rate of 33.9% in the export of high tech products, whereas their imports grew by 22.3% during the 1996-2001 period.

Figure 6.14. shows that during 1996-2001, the United States remains the leading exporter of high tech products, followed by the EU and Japan. However, Japan is the only one which showed a positive trade balance throughout the entire period. The high tech exports from Acceding Countries remain far below those of the United States, Japan or the EU. From 1996 to 2000, the global high tech export trends for the EU, Japan, the United States and Acceding Countries followed the same pattern as they all gradually increased. However, between 2000 and 2001 only EU-15 recorded an increase (4.3%) in high tech exports while Japan and the United States decreased by 7.3% and 20.5%, respectively.

The negative EU-15 high tech trade balance continued to increase from EUR 10.9 thousand million in 1996 to EUR 47.3 thousand million in 2000; then it decreased to only EUR 23.1 thousand million in 2001.

In 2001, the United States accounted for the largest share of world high tech exports (18.0%). High tech exports from the EU (excluding intra-EU trade) represented 15.0% of the world's high tech total exports. Following were Japan, Germany, France and the United Kingdom. The leading six countries together represented 57.4% of the world's high tech exports market share in 2001. During the 1996-2001 period, whilst the market share of high tech exports in the EU grew slowly (0.3% per annum), that of the United States and Japan fell on average each year by 1.8% and 6.9%, respectively. The rapid emergence of China as a high tech exporter is illustrated with it having the highest annual average growth rate of 17.5% between 1996 and 2001. Smaller economies like Mexico, Ireland, Hong Kong and the Netherlands experienced positive annual average growth rates – AAGR – at the expense of much larger economies – Figure 6.15.

Figure 6.13. **Annual average growth rates of high tech exports and imports in %**
EU-15, Acceding Countries, Japan and the United States
1996 to 2001

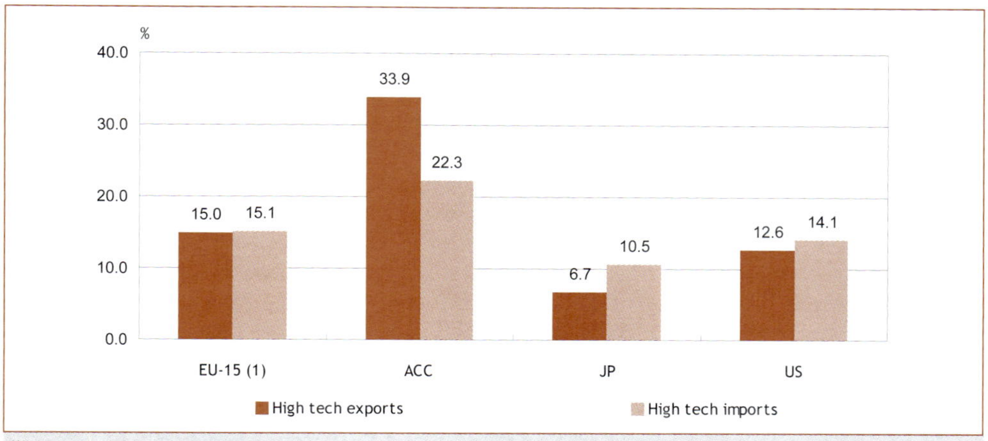

(1) EU-15 includes extra-EU trade only.

Sources: Eurostat, Comext; UN, Comtrade.

Figure 6.14. **Exports and trade balance for high tech products**
EU-15, Acceding Countries, Japan and the United States
1996 to 2001

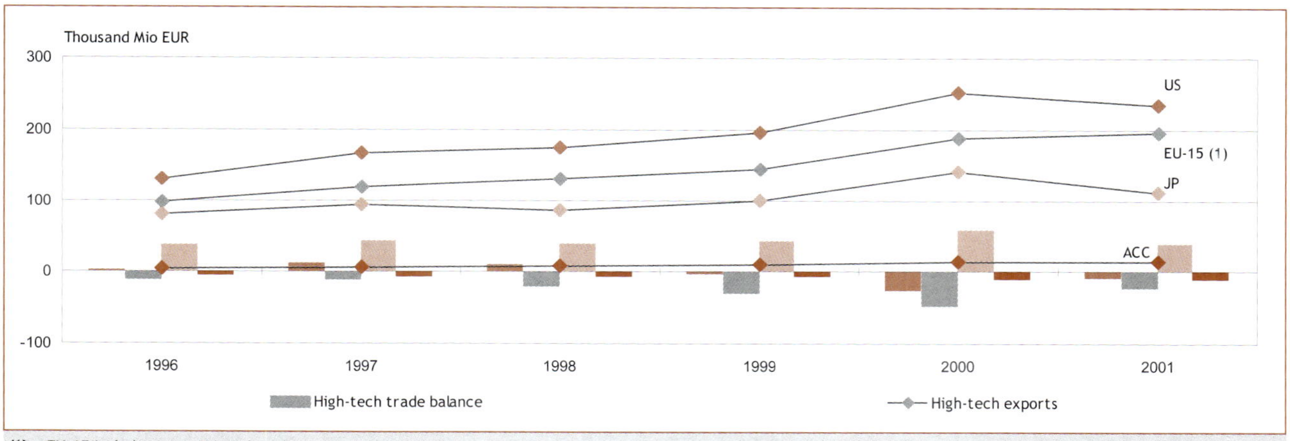

(1) EU-15 includes extra-EU trade only.

Sources: Eurostat, Comext; UN, Comtrade.

Figure 6.15. **High tech export world market shares in 2001 and their annual average growth rates in %**
in the world leading countries
1996 to 2001 (1)

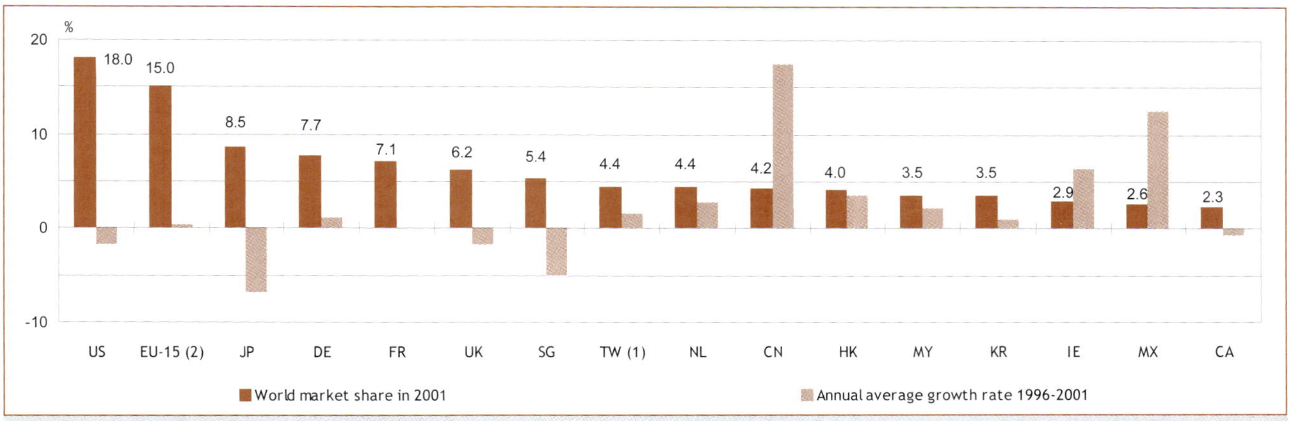

NB: The world total is estimated as the sum of 87 countries — see list in ***Methodological notes*** starting on page 150.
(1) The world total used to calculate market shares includes intra-EU trade.
(2) EU-15 includes extra-EU trade only, whereas individual Member States include both intra and extra-EU trade.

Sources: Eurostat, Comext; UN, Comtrade.

In terms of high tech trade performance, the strong position of the Philippines, Ireland, Malaysia and Singapore as net high tech exporters in 2001 is clearly shown in Figure 6.16., which ranks countries in terms of high tech trade balance as percentage of total exports. The EU's high tech trade balance has a percentage of total exports when only extra-EU trade is taken into account reveals a negative rate of -2.3%. Among the EU Member States, other countries with a significant trade surplus are Ireland, Finland, the United Kingdom, the Netherlands and France.

The extent to which a country's exports are specialised in the high tech sector relative to an average – in this case, the world's total exports – is shown by the relative specialisation index – RSI.

The relative specialisation index of a given country is defined as the proportion of high tech exports in the world's high tech exports, divided by the proportion of total exports in the world's total exports. Both the United States and Japan are specialised in high tech products, but their position has been surpassed by several countries in 2001. Philippines, Malta, Singapore and Malaysia are among the countries to have strengthened their position with an RSI value of 0.4 and above. Apart from Ireland, the United Kingdom, France, the Netherlands and Finland are the only other EU countries to be specialised in the exportation of high technology products – Figure 6.17.

Figure 6.16.

High tech trade balance as a percentage of total exports in the world leading countries 2001

(1) EU-15 includes extra-EU trade only.

Sources: Eurostat, Comext; UN, Comtrade.

Figure 6.17.

Relative specialisation index of high tech exports in the world leading countries 2001

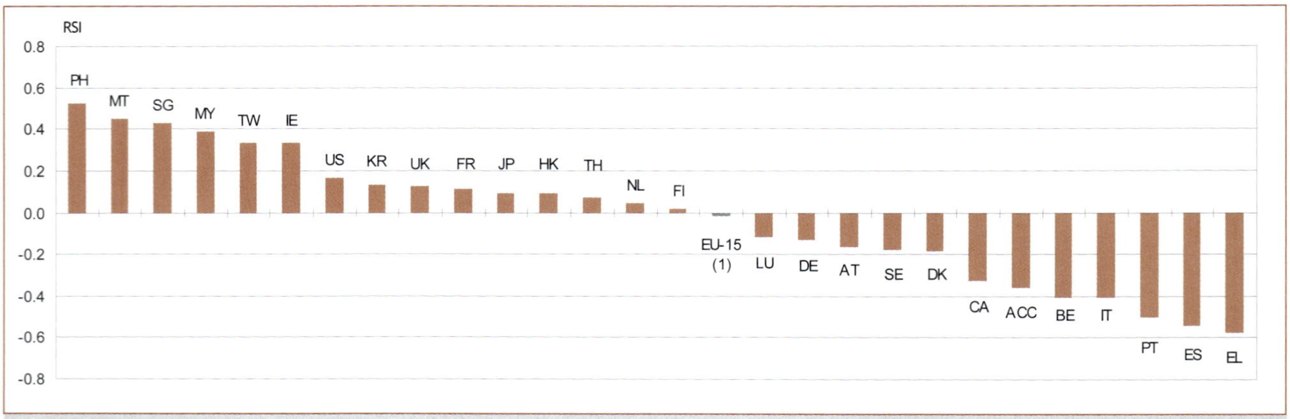

(1) EU-15 includes extra-EU trade only.

Sources: Eurostat, Comext; UN, Comtrade.

High tech trade in the EU

Whilst Germany leads in absolute terms, Ireland is the Member State most specialised in high tech trade

As shown in Table 6.13., high tech trade in Europe in 2001 accounted for about a fifth of its total trade. Note that this refers to EU high tech trade with the world but it excludes intra-EU trade. In value terms, high tech exports amounted to EUR 195.5 thousand million and the high tech imports reached EUR 218.6 thousand million, representing a negative balance of EUR 23.1 thousand million.

Considering the high tech trade of the individual EU countries, this time including trading with their EU partners, Germany, France and UK are the leading high tech traders in absolute terms. Ireland stood out given that around 40% of its trade is in high tech, compared to at most 28% for the remaining countries.

In terms of high tech exports in 2001, Germany led with EUR 100.7 thousand million representing 15.8% of its total trade. Next were France and the United Kingdom who exported EUR 92.4 and EUR 80.4 thousand million, respectively accounting for 25.6% and 26.4% of their total exports. Ireland exported only EUR 37.7 thousand million but that represented a high of 40.8% of its total exports.

With respect to high tech imports, Germany again leads in 2001 with EUR 98.8 thousand million, followed by France (EUR 86.6 thousand million) and the United Kingdom (EUR 72.5 thousand million). The lowest high tech trade values were reported for Greece and Luxembourg, although for the latter this represented over a quarter of its total trade.

Ireland had the highest high tech trade surplus of EUR 15.5 thousand million in 2001, followed by the United Kingdom (EUR 7.9 thousand million) and France (EUR 5.8 thousand million). Spain and Italy had the highest high tech trade deficit of over EUR 9 thousand million.

Table 6.13.

High tech exports, imports and trade balance
EU-15 (1)
2001

	High tech exports		High tech imports		High tech trade balance
	Thousand Mio EUR	As a % of total exports	Thousand Mio EUR	As a % of total imports	Thousand Mio EUR
EU-15 (2)	195.5	19.8	218.6	21.3	-23.1
BE	19.1	9.0	21.6	10.8	-2.5
DK	8.1	14.0	7.8	15.4	0.3
DE	100.7	15.8	98.8	18.2	1.9
EL	0.6	5.5	3.2	10.1	-2.5
ES	7.9	6.1	18.2	10.6	-10.3
FR	92.4	25.6	86.6	23.6	5.8
IE	37.7	40.8	22.2	39.3	15.5
IT	23.2	8.6	32.2	12.4	-9.0
LU	3.0	27.9	3.6	25.8	-0.5
NL	57.4	22.3	52.8	22.7	4.7
AT	11.6	14.6	12.6	15.1	-1.0
PT	1.9	6.8	5.0	11.4	-3.2
FI	10.2	21.1	6.6	18.2	3.6
SE	12.0	14.2	11.5	16.3	0.5
UK	80.4	26.4	72.5	19.5	7.9

(1) All figures for individual countries include both intra and extra EU trade.
(2) EU-15 includes extra-EU trade only.

Source: Eurostat, Comext.

Extra-EU trade of high tech products

41.9% of the EU's total high tech exports in 2001 went to non EU countries

Table 6.14. shows the distribution of EU high tech trade flows by main partners for the 1999-2001 period. It can be seen that intra-EU high tech exports accounted for 59.3% of total EU high tech exports. Excluding intra-EU flows, a total of EUR 195.5 thousand million of high tech products were exported from the EU in 2001. 28.8% of these products were exported to the United States and 4.2% to Japan. With regard to imports, just over half of the total EU high tech imports originate from other Member States. Extra-EU high tech imports alone accounted for EUR 218.6 thousand million in 2001, of which 35.3% were imported from the United States, and 10.5% from Japan.

As shown in Figure 6.18., the EU high tech trade balance was in deficit between 1999 and 2001, and fluctuated from EUR 31 thousand million in 1999 up to EUR 47 thousand million in 2000 and down by half to EUR 23 thousand million in 2001.

Among the EU Member States, the countries with the highest extra-EU high tech trade deficits (excluding intra-EU trade) are the Netherlands, the United Kingdom and Germany with EUR 25 thousand million, EUR 9 thousand million and EUR 4 thousand million each in 2001, respectively. France has the highest high tech trade surplus (EUR 10 thousand million in 2001) followed by Sweden (EUR 4 thousand million) and Finland (EUR 3 thousand million).

Table 6.14. **Distribution of high tech exports and imports by selected partner EU-15 1999 to 2001**

		High tech exports			High tech imports		
		1999	2000	2001	1999	2000	2001
Total in thousand Mio EUR (1)		353.4	455.9	466.3	360.7	471.9	455.2
Of which by partner in %	Intra EU	59.3	58.9	58.1	51.6	50.3	52.0
	JP	1.7	1.9	1.8	5.8	6.0	5.0
	US	11.3	11.4	12.1	18.2	17.6	17.0
	Other	27.7	27.9	28.1	24.4	26.1	26.0
Extra EU in thousand Mio EUR		143.9	187.4	195.5	174.5	234.7	218.6
Of which by partner in %	JP	4.1	4.6	4.2	12.1	12.0	10.5
	US	27.8	27.6	28.8	37.5	35.4	35.3
	Other	68.1	67.8	67.0	50.4	52.6	54.2

(1) Total includes both intra and extra EU trade.

Sources: Eurostat, Comext; UN, Comtrade.

Figure 6.18. **Extra-EU high tech trade balance EU-15 1999 to 2001**

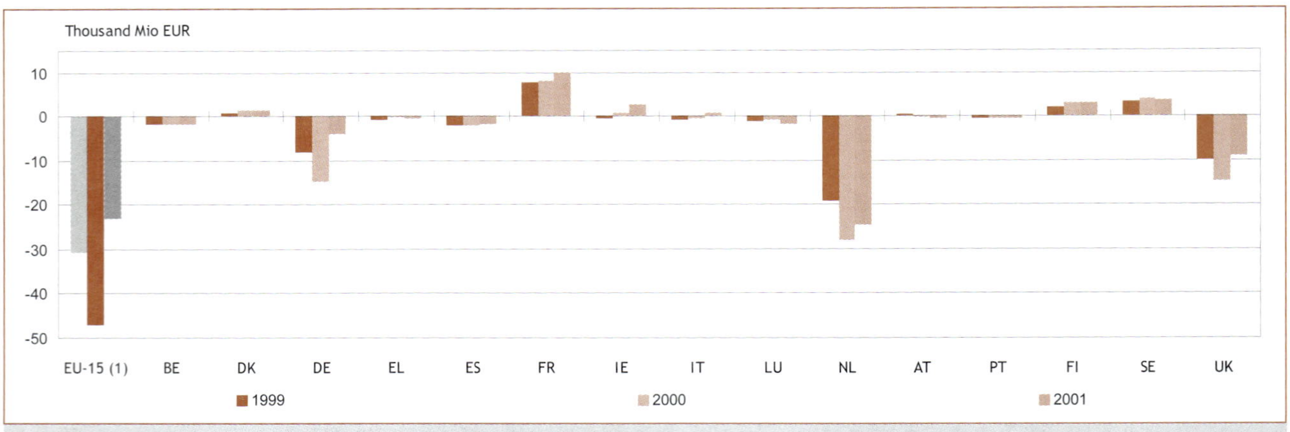

(1) EU-15 aggregate as well as data for EU Member States include extra-EU trade only.

Sources: Eurostat, Comext; UN, Comtrade.

Table 6.15. **High tech exports and imports EU-15, Candidate Countries, Iceland, Norway, Japan and the United States 2001**

	High tech exports		High tech imports	
	Thousand Mio EUR	As a % of total exports	Thousand Mio EUR	As a % of total imports
EU-15 (1)	195.5	19.8	218.6	21.3
BE	5.4	10.2	7.0	11.6
DK	3.4	17.4	2.1	12.9
DE	48.1	16.8	52.3	21.5
EL	0.5	6.8	1.0	7.0
ES	2.7	7.3	4.6	8.1
FR	45.8	32.4	35.8	28.0
IE	12.7	37.2	9.9	51.3
IT	11.9	9.5	11.1	9.8
LU	0.2	13.3	1.9	66.5
NL	13.0	23.7	37.6	33.4
AT	5.1	16.7	5.5	20.7
PT	0.8	15.1	1.5	13.3
FI	5.5	24.8	2.5	18.7
SE	7.8	20.3	4.2	17.3
UK	32.4	25.1	41.6	22.4
ACC	14.5	9.7	25.8	13.8
CZ	3.4	9.2	6.1	15.0
EE	0.7	14.6	0.6	11.0
CY	0.0	1.5	0.5	10.9
LV	0.1	2.2	0.3	8.5
LT	0.1	2.9	0.6	7.8
HU	7.0	20.7	7.7	20.5
MT	1.2	59.4	1.1	34.7
PL	1.0	2.6	6.5	11.6
SI	0.5	4.8	0.9	8.2
SK	0.5	3.7	1.5	9.1
BG (2)	0.1	1.6	0.6	8.3
RO	0.6	5.0	1.9	10.7
TR	1.1	3.2	5.4	11.6
IS	0.0	1.3	0.1	12.7
NO	2.6	3.9	3.2	15.8
JP	111.2	24.7	72.0	18.5
US	233.8	28.6	243.3	18.5

(1) EU-15 aggregate as well as data for EU MS include extra-EU trade only.
(2) **Exception to the reference year 2001**
BG: 2000.

Sources: Eurostat, Comext; UN, Comtrade.

In 2001, the United States had the highest percentage (28.6%) of its total exports accounted for by high tech products – Table 6.15. This was followed by Japan (24.7%), the EU (19.8%) and the Acceding Countries aggregate (9.7%). In terms of extra-EU trade at the EU Member State level, high tech products in Ireland and France accounted for the highest proportions, as they amounted to 37.2% and 32.4% of their total exports, respectively. The United Kingdom, Finland and the Netherlands followed with around a quarter of their total exports being high tech products.

Examining the high tech component of total imports in 2001, the EU has the highest proportion (21.3%) of high tech goods, but very closely followed by Japan and the United States both with 18.5% of their respective total imports. 13.8% of the goods imported by Acceding Countries were high tech products.

Among the EU countries and considering extra-EU trade only, high tech imports in Luxembourg accounted for almost two-thirds of its total imports in 2001. Over half of the total imports in Ireland and a third in the Netherlands were high tech products that year, whereas 28% of French imports were accounted for by high tech products.

Figure 6.19. shows that between 1996 and 2001, high tech trade grew faster than total trade in the EU, which retained the highest high tech export growth with an annual average growth rate of 15.0%. Whilst the United States grew at 12.6% per annum, high tech exports from Japan increased at a rate of 6.7%. In Japan high tech exports and imports grew below their respective totals. While high tech exports increased above total exports in the United States, the growth of imports of high tech products was below that of total imports.

Figure 6.19. **Annual average growth rates of total and high tech exports and imports in % EU-15, Japan and the United States 1996 to 2001**

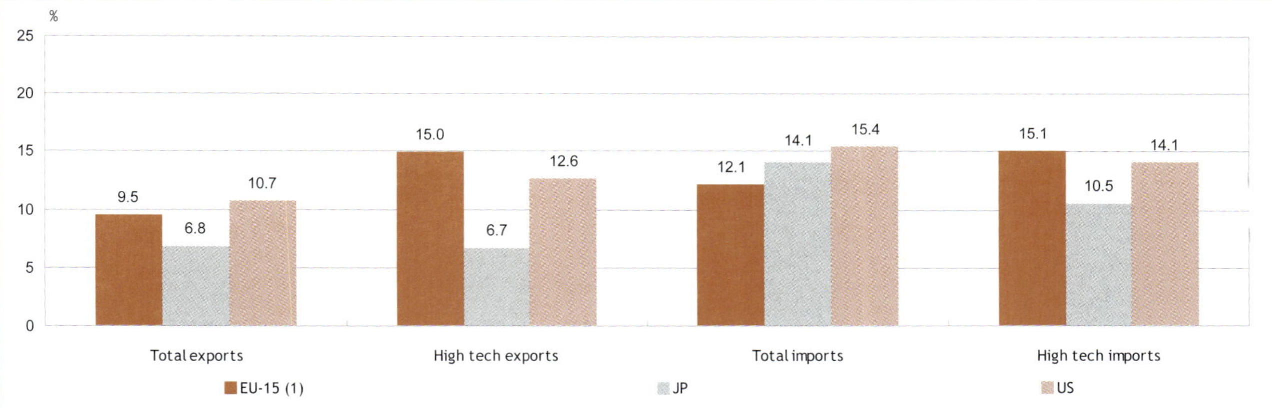

(1) EU-15 includes extra-EU trade only.

Sources: Eurostat, Comext; UN, Comtrade.

Intra-EU trade of high tech products

Germany is the main export partner of high tech products in the EU

In terms of exports of high tech trade products within the EU in 2001, the three main export partners of the Member States were Germany, the United Kingdom and France – Table 6.16. Germany ranks first as it is the principal export partner for 9 countries and the second main export partner for another 5 countries. The second high tech exporter within the EU is the United Kingdom, which is the principal partner for three Member States and second biggest partner for six others. France is the third largest export partner of EU countries after Germany and the United Kingdom.

With regard to imports of high tech trade products within the EU in 2001 – Table 6.17., again Germany remains the largest supplier of high tech products to the Member States, followed by the United Kingdom. However, Netherlands is the third principal supplier. Neighbouring Belgium and Germany import 22.7% and 27.9% of high tech products from Netherlands, respectively. The Netherlands is also the second biggest supplier of high tech products to the United Kingdom (23.3%), Italy (19.2%), Greece (18.0%) and Austria (15.3%). France is the fourth largest import partner of EU countries. The highest intra-EU import ratio for high tech products was 68.0%, which represented the proportion of high tech imports by Ireland from the United Kingdom.

Table 6.16.

Distribution of intra-EU high tech exports
in %
2001

Reporting country	Partner country: BE	DK	DE	EL	ES	FR	IE	IT	LU	NL	AT	PT	FI	SE	UK	EU-15
BE	:	1.4	20.4	1.4	5.3	23.5	1.8	9.7	4.3	12.3	2.3	1.1	1.0	2.7	12.9	100
DK	2.1	:	28.4	0.8	3.4	7.0	4.0	5.1	2.9	8.4	1.3	0.7	4.3	15.1	16.4	100
DE	4.7	2.2	:	1.2	7.1	29.2	1.8	11.4	0.7	7.5	6.6	2.3	2.9	3.2	19.1	100
EL	2.5	5.6	16.1	:	0.3	14.7	9.3	6.2	0.0	13.5	0.7	0.4	0.2	2.1	28.4	100
ES	3.6	0.9	21.5	0.6	:	19.2	1.4	8.6	0.5	6.8	3.4	19.0	0.6	1.6	12.3	100
FR	7.9	1.2	30.5	0.6	8.0	:	3.4	10.3	2.5	6.0	1.4	6.9	1.5	3.0	16.8	100
IE	3.8	1.4	17.7	0.2	3.4	9.4	:	5.1	0.0	10.1	0.9	0.4	0.9	3.4	43.2	100
IT	17.5	1.3	19.4	2.7	8.2	23.6	3.9	:	1.6	5.3	2.9	1.4	0.7	1.6	9.8	100
LU	8.9	0.8	21.4	0.6	6.4	19.3	3.6	10.2	:	2.5	0.7	1.4	1.7	2.0	20.7	100
NL	6.3	2.6	27.6	1.0	6.6	14.8	2.0	10.2	0.5	:	1.9	1.4	2.1	3.7	19.3	100
AT	2.6	1.0	44.7	1.2	4.6	12.8	0.8	8.5	0.8	5.5	:	1.5	1.1	2.8	12.0	100
PT	3.2	1.0	54.2	0.3	18.8	5.8	0.4	2.6	0.0	7.7	0.5	:	0.1	0.8	4.6	100
FI	2.3	4.0	16.8	3.4	4.1	11.3	2.5	8.0	0.1	5.1	3.1	1.7	:	9.2	28.5	100
SE	5.4	9.8	21.4	2.2	5.9	9.6	1.6	10.2	0.1	6.2	2.3	1.5	9.0	:	14.9	100
UK	3.4	2.8	20.2	0.7	6.7	19.6	17.0	7.1	0.6	14.5	1.4	1.0	1.5	3.5	:	100

Source: Eurostat, Comext.

Table 6.17.

Distribution of intra-EU high tech imports
in %
2001

Reporting country	Partner country: BE	DK	DE	EL	ES	FR	IE	IT	LU	NL	AT	PT	FI	SE	UK	EU-15
BE	:	1.0	17.1	0.3	1.5	18.2	7.4	14.4	1.2	22.7	1.1	0.2	0.8	2.3	11.9	100
DK	2.3	:	22.8	0.2	1.0	10.9	6.1	2.3	0.3	15.8	0.8	0.2	3.9	10.4	23.0	100
DE	5.1	2.2	:	0.1	2.2	21.9	8.9	4.0	0.6	27.9	4.6	1.5	0.7	1.7	18.5	100
EL	4.5	0.5	27.8	:	1.2	8.9	1.9	9.2	0.6	18.0	0.7	0.2	6.6	9.0	10.9	100
ES	4.0	1.1	23.6	0.0	:	19.2	5.6	7.2	0.7	17.0	2.0	1.5	1.4	2.6	14.2	100
FR	7.7	1.2	33.2	0.3	2.8	:	4.6	7.1	2.5	14.2	2.0	5.9	1.3	1.2	16.0	100
IE	1.7	0.9	8.4	0.1	0.4	9.8	:	1.2	0.0	7.4	0.4	0.0	1.0	0.7	68.0	100
IT	5.1	1.0	27.7	0.2	1.9	16.5	5.6	:	1.2	19.2	2.1	0.3	1.7	3.1	14.4	100
LU	28.5	0.2	15.4	0.0	8.6	12.3	1.0	1.1	:	9.4	2.5	0.2	0.0	2.0	18.6	100
NL	4.7	1.2	22.5	0.1	2.7	10.2	13.4	3.9	0.9	:	1.8	0.4	2.0	2.3	33.9	100
AT	3.1	0.7	51.5	0.0	2.1	6.3	2.9	4.6	0.2	15.3	:	0.1	4.1	1.7	7.3	100
PT	3.5	0.6	26.1	0.0	24.5	10.2	2.0	3.6	0.5	14.8	2.7	:	1.3	3.6	6.6	100
FI	2.0	3.7	27.9	0.0	1.7	14.1	7.7	3.8	0.0	7.5	1.3	0.4	:	9.5	20.3	100
SE	3.6	6.0	21.6	0.0	1.2	12.6	10.2	1.5	0.5	17.4	1.6	0.2	4.6	:	19.0	100
UK	4.3	1.5	23.6	0.2	2.2	13.9	16.3	3.1	0.7	23.3	2.6	0.2	4.9	3.3	:	100

Source: Eurostat, Comext.

High tech trade flows in the Acceding Countries

45.5% of high tech exports from Acceding Countries went to the EU

Between 1996 and 2001, the high tech exports from Acceding Countries quadrupled in value from EUR 3.4 thousand million to EUR 14.5 thousand million, while its high tech imports tripled from EUR 9.4 thousand million to EUR 25.8 thousand million – Table 6.18. The EU was by far the largest high tech trade partner of the Acceding Countries. In 1996, Acceding Countries exported EUR 1.5 thousand million of high tech products to the EU, which represented 45.9% of its total high tech exports. By 2001, high tech exports rose to EUR 10.2 thousand million, 70.0% of its high tech exports. High tech exports of Acceding Countries grew at an annual average growth rate of 33.9% during the 1996-2001 period. During this period, exports of Acceding Countries to the EU and the United States grew at annual average growth rates of 45.5% and 34.9%, respectively. High tech imports with these two large partners grew slower, registering annual average growth rates of 18.9% with the EU and 20.2% with the United States.

Between 1995 and 2001, the high increase of high tech exports from Acceding Countries was largely determined by the increase in its exports to the EU. However, throughout this period, the Acceding Countries experienced a trade deficit, which more than doubled from nearly EUR 5 thousand million in 1995 to EUR 11 thousand million in 2001 – Figure 6.20. This deficit arises mainly from a great increase in imports from other countries, which grew at annual average growth rates of 30.3%. The main supplier of high tech products for the Acceding Countries is the EU, which accounted for 49.3% of the high tech imports in 2001.

Table 6.18. **Distribution of high tech trade by main partner Acceding Countries 1996 to 2001**

High tech exports					High tech imports				
Partners		1996	2001	AAGR in % 1996-2001	Partners		1996	2001	AAGR in % 1996-2001
Total in thousand Mio EUR		3.4	14.5	33.9	Total in thousand Mio EUR		9.4	25.8	22.3
Of which by partner in %	EU-15	45.9	70.0	45.5	Of which by partner in %	EU-15	57.0	49.3	18.9
	ACC	12.4	4.7	10.5		ACC	3.5	2.9	17.8
	JP	1.6	0.9	19.3		JP	4.8	5.6	26.2
	US	7.6	8.0	34.9		US	12.0	10.9	20.2
	Others	32.5	16.3	18.5		Others	22.7	31.2	30.3

Sources: Eurostat, Comext; UN, Comtrade.

Figure 6.20. **High tech trade exports and balance by selected partner Acceding Countries 1995 to 2001**

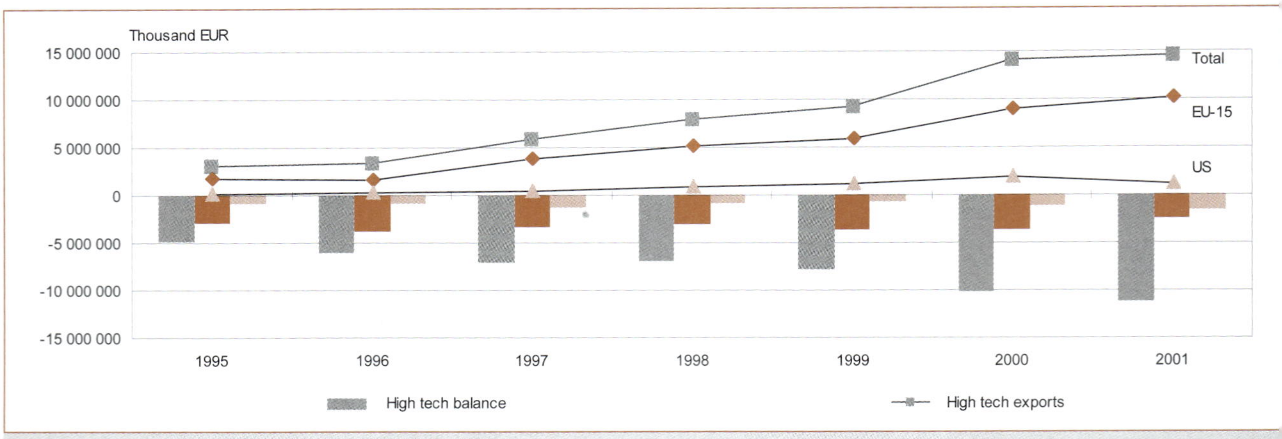

Sources: Eurostat, Comext; UN, Comtrade.

Distribution by product group

'Electronics' account for the highest proportion of high tech trade in the EU, Acceding Countries, Japan and the United States

Looking at the distribution of high tech trade by product group in Table 6.19., 'Electronics' are by far the most traded goods, followed by 'Computer & office machinery' products and 'Aerospace'.

'Electronics' contribute to at least 30% of high tech exports and imports in the EU, Acceding Countries, Japan and the United States. The next highest export categories are 'Aerospace' products in the EU and the United States, and 'Computer & office machinery' in the Acceding Countries and Japan.

With regards to high tech imports, 'Computer & office machinery' makes up the second largest category in the EU, Acceding Countries, the United States and Japan, as this group contributes to at least a quarter to a third of their high tech imports.

Table 6.19.

High tech trade exports and imports in % by product group EU-15, Acceding Countries, Japan and the United States 2001

	High tech exports in %				High tech imports in %			
	EU-15 (1)	ACC	JP	US	EU-15 (1)	ACC	JP	US
Aerospace	25.0	3.4	1.1	20.6	17.8	2.6	5.1	11.9
Armement	0.7	1.3	0.1	1.3	0.3	0.9	0.4	0.4
Chemicals	3.3	1.7	1.0	2.1	2.2	4.8	3.8	2.0
Computers & office machinery	13.8	28.1	24.2	18.3	26.5	25.2	31.6	31.1
Electrical machinery	2.4	5.4	6.0	1.9	3.4	4.9	4.0	2.3
Electronics	30.9	47.2	48.4	35.4	32.5	43.4	37.7	37.3
Instruments	11.8	6.1	12.9	13.3	9.5	8.0	11.4	9.2
Non electrical machinery	4.6	4.1	5.3	3.9	3.4	4.5	2.6	2.8
Pharmaceuticals	7.3	2.7	1.0	3.1	4.4	5.5	3.5	3.0
Total high tech in thousand Mio EUR	195.5	14.5	111.2	233.8	218.6	25.8	72.0	243.3

(1) EU-15 includes extra-EU trade only.

Sources: Eurostat, Comext; UN, Comtrade.

Examining the annual average growth rates – AAGR – of total high tech exports in Figure 6.21., the Acceding Countries have experienced the highest growth rates (33.9%) between 1996 and 2001, compared to EU (15.0%), Japan (6.7%) and the United States (12.6%). For the Acceding Countries, high growth rates of exports were observed in 'Computers and office machinery' (69.0%), and 'Electronics' (36.0%). Growth rates were more consistent and lower for the high tech product groups in EU, Japan and the United States. In the EU, the annual average growth rates were highest in 'Pharmaceuticals', 'Electronics', 'Aerospace' and 'Electrical machinery'.

As shown in Figure 6.22. with respect to high tech imports, the Acceding Countries again had the highest annual average growth rate of 22.3% between 1996 and 2001. This was followed by the EU, the United States and Japan. Product groups with the highest annual average growth rates for imports in the Acceding Countries were 'Electrical machinery', 'Non electrical machinery' and 'Computers and office machinery'. In the case of the United States, 'Aerospace' products had the highest rate which mounted to 30.0%. The EU observed the highest annual average growth rates in the imports of 'Electronics' and 'Aerospace'.

Figure 6.21.

**Annual average growth rates of high tech exports by product group in %
EU-15, Acceding Countries, Japan and the United States
1996 to 2001**

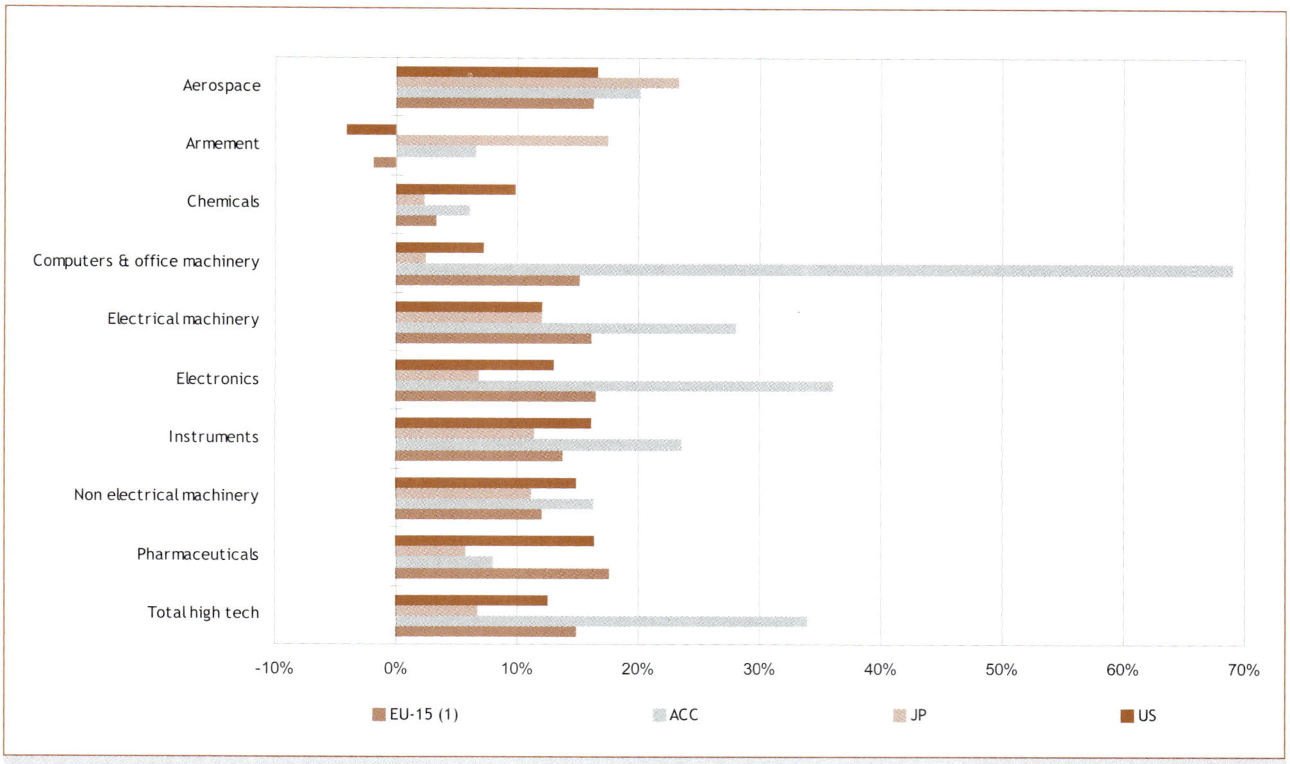

([1]) EU-15 includes extra-EU trade only.

Sources: Eurostat, Comext; UN, Comtrade.

Figure 6.22.

**Annual average growth rates of high tech imports by product group in %
EU-15, Acceding Countries, Japan and the United States
1996 to 2001**

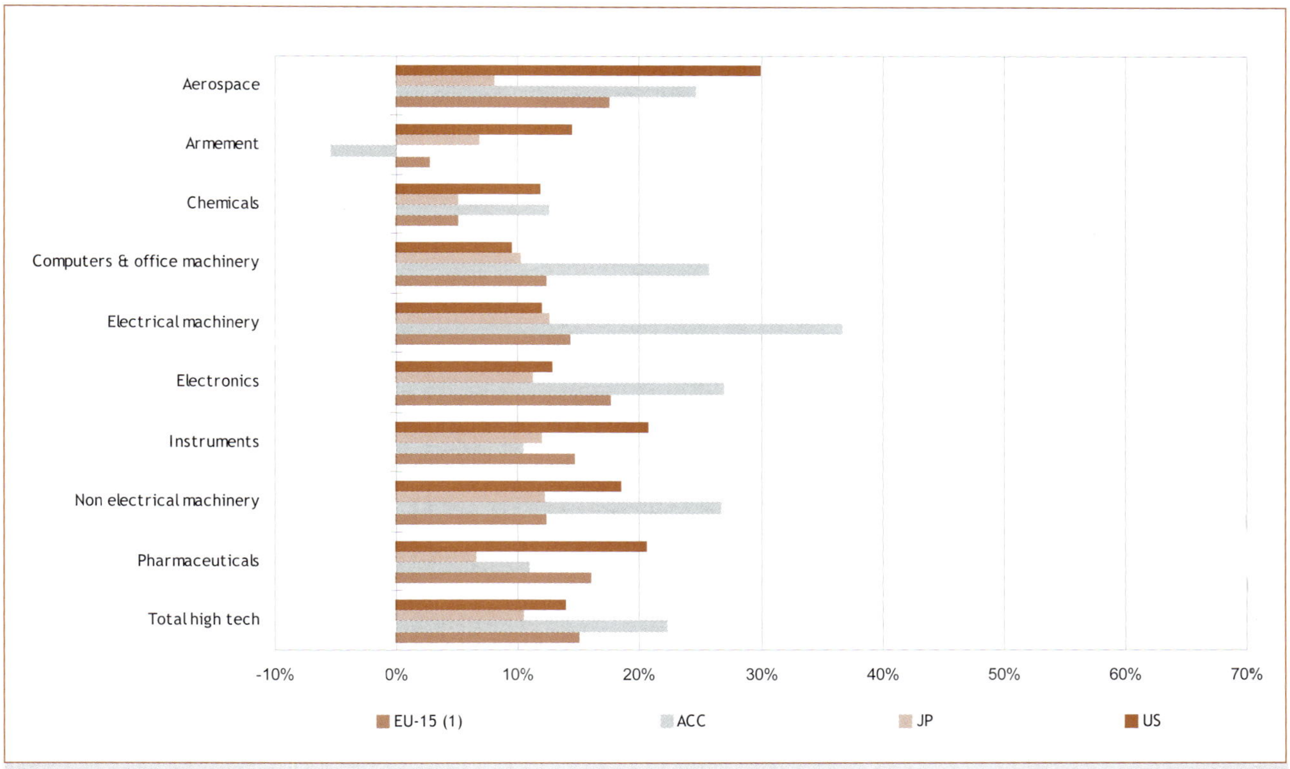

([1]) EU-15 includes extra-EU trade only.

Sources: Eurostat, Comext; UN, Comtrade.

Methodological notes

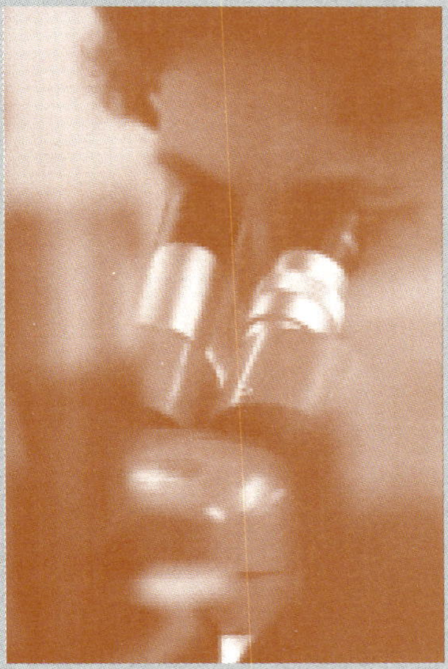

This part presents in some detail, the methodology used for the data presented in this publication. After giving some general information, specific details are given for the following domains: Government budget appropriations or outlays for R&D – GBAORD, R&D expenditure and personnel, Human Resources in Science and Technology – HRST, Patents, employment in high technology sectors, value added and external trade of high tech products.

1. General information

Currencies

Series in current EUR, have been calculated by using current exchange rates.

Data measured in constant 1995 Purchasing Power Standard – PPS – are first corrected for inflation using the GDP deflator – *a Paasche index with 1995 = 100 as a base* – of the country in question before applying the 1995 PPS exchange rate.

Regional data

Regional data presented throughout this publication are treated at Eurostat according to the guidelines established by *The Regional Dimension of R&D and Innovation Statistics and Experimental Development – Regional Manual*, European Commission, 1996.

Nomenclature of territorial units for statistics – NUTS

The regional data presented in this publication are broken down according to the *Nomenclature of Territorial Units for Statistics* – NUTS – classification, 1998 version. The NUTS was established by the Statistical Office of the European Communities, in co-operation with the Commission's other departments, to provide a single, uniform breakdown of territorial units for the production of regional statistics for the European Union.

The NUTS is a five-level hierarchical classification comprising three regional and two local levels. In this way, NUTS subdivides each Member State into a whole number of NUTS 1 regions, each of which is in turn subdivided into a whole number of NUTS 2 regions, and so on. In the present publication most data are presented at NUTS 2 level on the basis of the NUTS 1998 version. The exceptions have been indicated in the tables or figures.

For Denmark and Luxembourg the national level coincides with the NUTS 2 level, which explains their potential presence amongst the regional rankings in this publication.

Iceland and Norway are not included in the NUTS classification but do have similar statistical regions. As for Denmark and Luxembourg, Iceland is also classified at the statistical region level 2.

Acceding Countries – ACC – and Candidate Countries

The Acceding Countries aggregate – ACC – comprises Cyprus, the Czech Republic, Estonia, Hungary, Latvia, Lithuania, Malta, Poland, the Slovak Republic and Slovenia. Exceptions to the countries included are specified in the corresponding tables or figures.

The term Candidate Countries refers to all Acceding Countries, Bulgaria, Romania and Turkey.

Annual average growth rates – AAGR

Annual average growth rates in this publication are calculated according to the following formula:

$$AAGR = \left(\left[(X_n - X_{n-t})/X_{n-t}\right]^{1/t} - 1\right) \times 100$$

2. Specific methodological notes by domain

Government budget appropriations or outlays for R&D – GBAORD

Definition

Government budget appropriations or outlays on R&D are all appropriations allocated to R&D in central government or federal budgets and therefore refer to budget provisions, not to actual expenditure. Provincial or state government should be included where the contribution is significant. Unless otherwise stated, data include both current and capital expenditure and cover not only government-financed R&D performed in government establishments, but also government-financed R&D in the business enterprise, private non-profit and higher education sectors, as well as abroad (i.e. international organisations). Data on actual R&D expenditure, which are not available in their final form until some time after the end of the budget year concerned, may well differ from the original budget provisions. This and further methodological information can be found in the *Frascati Manual*, OECD, 2002.

GBAORD data do not consider the amount of money actually spent, but are rather based on budget provisions, and so should be seen as intentions of spending. These data reflect policies at a given moment in time and the concomitant priorities of the policy makers when allocating their budgets. These data are hard to collect because they are not obtained from *ad-hoc* surveys, but in most cases are obtained from national budget statistics. The difficulty is due more specifically to the fact that national budgets already have their own terminology and methodology and therefore do not accord entirely with the Eurostat guidelines and the methodology proposed by the Frascati Manual.

Data are collected at the national level and the procedure can be articulated in a two step process:

- within the budget statistics, it is first necessary to identify the budget items that involve R&D;
- the R&D content of these budget items must then be measured or estimated.

Methodological discrepancies and exceptions

Despite all efforts, the concepts and methods used by the individual Member States of the EU-15, the United States and Japan for collecting data on government R&D appropriations are not completely harmonised.

Details on each country's methodology can be found in:

- *Statistics on Science and Technology: Annual Statistics – data 1991-2001* or in
- Eurostat's reference database *NewCronos*, Theme 9, Domain *GBAORD*.

No GBAORD data exist for Luxembourg before 2000 and therefore EU-15 totals exclude Luxembourg before that year. EU-15 totals include Luxembourg for the years from 2000 onwards unless otherwise stated in the chapter.

United States data exclude the socio-economic objectives 'Research financed from General University Funds' and 'Other civil research' and are therefore systematically underestimated. Comparisons with other countries should be made with caution.

The figures for Japan are estimates made by the OECD Secretariat and recognised as official data by the Japanese Government. They underestimate expenditure on the social and human sciences and are thus only to some extent comparable with the data for other countries. Moreover, data are in general underestimated because the R&D portion of military contracts is excluded.

Breakdown by socio-economic objectives — NABS

Government R&D appropriations are broken down by socio-economic objective on the basis of the *Nomenclature for the analysis and comparison of scientific programmes and budgets* — NABS, Eurostat, 1994. The 1983 version of NABS applies to all the figures up until the 1992 final budgets and the 1993 provisional budgets. The 1993 version applies from the 1993 final and the 1994 provisional budgets onwards. As a result of the revision of NABS, some caution should be employed when comparing the data for some NABS headings with those of earlier years.

The greatest differences are to be found in the following chapters of the NABS:

- Chapter 1 Exploration and exploitation of the earth;
- Chapter 3 Control and care of the environment;
- Chapter 5 Production, distribution and rational utilisation of energy;
- Chapter 7 Industrial production and technology;
- Chapter 10 Research financed from General University Funds (GUF);
- Chapter 11 Non-oriented research.

Furthermore, not all countries transpose their data directly to NABS: some follow other compatible classifications — OECD, Nordforsk — which are then converted to the NABS classification, Table 8.2. of the *Frascati Manual*, 2002.

For a more user friendly presentation, socio-economic objectives may be grouped in various categories, which are defined as follows:

Grouped socio-economic objectives	NABS chapters
• Research financed from General University Funds (GUF)	10
• Technological objectives	1 + 5 + 7 + 9
• Defence	13
• Non-oriented research	11
• Human and social objectives	2 + 3 + 4 + 8
• Agricultural production and technology	6
• Other civil research	12

For further information on the breakdown by socio-economic objectives of GBAORD data, please refer to Chapter 8.7. of the *Frascati Manual*, OECD 2002.

Time series

The analysis in Chapter 1 covers the period 1992 to 2002, with 2002 data being provisional.

Sources

- Eurostat, *NewCronos*; Theme 9, Domain *GBAORD* and Domain *RD_CEC*;
- United States and Japan: OECD, *Main Science and Technology indicators* — MSTI 2002/2.

National reports on their specific GBAORD performance are available for various countries at Eurostat. For these reports and further information on definitions and explanatory notes, see:

- Metadata in Eurostat's reference database *NewCronos*, Theme 9, Domain GBAORD or in
- *Statistics on Science and Technology: annual statistics — Data 1991-2001*.

R&D expenditure and personnel

The basic concepts, guidelines for collecting data and the classifications to be used in compiling statistics on research and experimental development are given in the *Frascati Manual* (1). Regional data are collected according to the standards defined by the *Regional Manual* (2)

Research and experimental development — R&D

Research and experimental development (R&D) activities comprise creative work undertaken on a systematic basis in order to increase the stock of knowledge, including knowledge of man, culture and society and the use of this stock of knowledge to devise new applications. There are two basic statistical variables in this domain, namely R&D expenditure and R&D personnel.

R&D indicators for R&D expenditure

R&D expenditure corresponds to the measurement of 'intramural' expenditure, i.e. all expenditure for R&D performed within a statistical unit or sector of the economy during a specific period, whatever the source of funds (3).

- R&D intensity

R&D intensity represents the R&D expenditure as a percentage of GDP. It is the ratio of R&D expenditure in current EUR for the sectors and years in question to GDP.

- Fields of science

Data on R&D expenditure may be broken down by field of science. The classification of fields of science is based on the nomenclature suggested by Unesco: *Recommendation concerning the International Standardisation of Statistics on Science and Technology* — See the *Frascati Manual*, Sections 4.4., 3.6.2. and 3.7.2.

R&D indicators for R&D personnel

All persons employed directly on R&D should be counted, as well as those providing direct services such as R&D managers, administrators and clerical staff. Those providing indirect services, such as canteen and security staff, should be excluded — *Frascati Manual*, § 294-296.

- Researchers

Researchers are professionals engaged in the conception or creation of new knowledge, products, processes, methods and systems, and in the management of the projects concerned — *Frascati Manual*, § 301.

- Full-time equivalent — FTE

One FTE may be thought of as one person-year. For instance, a person who normally spends 40% of his time on R&D and the rest of it on other work (e.g. lecturing, university administration, guidance) should be counted as only 0.4 FTE — *Frascati Manual*, Section 5.3.3.

(1) ***Standard method for surveys on R&D and experimental development — Frascati Manual,*** OECD 2002.

(2) ***The regional dimension of R&D statistics and of innovation — Regional Manual,*** Eurostat, *1996.*

(3) ***Standard method for surveys on R&D and experimental development — Frascati Manual,*** OECD 2002, § 358.

- Personnel by number of individuals – HC

The number of individuals who are employed mainly or partly on R&D – *Frascati Manual*, Section 5.3.2.

- Labour force

The labour force is the active population, this is the sum of employed and unemployed persons as defined by the EU Labour Force Survey. Persons in employment are those who during the reference week did any work for pay or profit, or were not working but had jobs from which they were temporarily absent, including family workers. Unemployed persons comprise persons aged 15 to 74 who were:

- without work during the reference week, i.e. neither had a job nor were at work (for one hour or more) in paid employment or self-employment;
- currently available for work, i.e. were available for paid employment or self-employment before the end of the two weeks following the reference week;
- actively seeking work, i.e. had taken specific steps in the four weeks period ending with the reference week to seek paid employment or self-employment or who found a job to start later, i.e. within a period of at most three months.

Institutional classifications

Intramural R&D expenditure and R&D personnel may be broken down with reference to the four institutional sectors in which the R&D takes place.

- The business enterprise sector – BES

With regard to R&D, the business enterprise sector includes: all firms, organisations and institutions whose primary activity is the market production of goods or services (other than higher education) for sale to the general public at an economically significant price and the private non-profit institutions mainly serving them – *Frascati Manual*, § 163.

- The government sector – GOV

In the field of R&D, the government sector includes: all departments, offices and other bodies which furnish but normally do not sell to the community those common services, other than higher education, which cannot otherwise be conveniently and economically provided, and administer the state and the economic and social policy of the community (public enterprises are included in the business enterprise sector) as well as PNPs controlled and mainly financed by government – *Frascati Manual*, § 184.

- The higher education sector – HES

This sector comprises: all universities, colleges of technology and other institutes of post-secondary education, whatever their source of finance or legal status. It also includes all research institutes, experimental stations and clinics operating under the direct control of or administered by or associated with higher education establishments – *Frascati Manual*, § 206.

- The private non-profit sector – PNP

This sector covers: non-market, private non-profit institutions serving households (i.e. the general public) and private individuals or households – *Frascati Manual*, § 194.

Nomenclature of territorial units for statistics – NUTS

In chapters 2 and 3 of the present Panorama, regional data are presented at NUTS 2 level on the basis of the NUTS 1998 version. The exceptions have been indicated in the tables or figures.

Data for United Kingdom are only available at NUTS 1 level.

European aggregates

For both R&D expenditure and personnel, EU totals are calculated as the sum of the national data by sector. If data are missing, estimates are first made for the country in question, reference period, institutional sector or relevant R&D variable, as appropriate. This method is not identically applied to the calculation of R&D personnel in head count (HC). The estimates for R&D personnel in full time equivalents (FTE) serve as a basis for the HC calculation. An FTE/HC ratio based on available FTE and HC personnel data at the national level is estimated for the EU aggregates, by institutional sector and by year. This ratio is then applied to the FTE data to calculate the EU totals in HC.

- EU-15 and EEA data are estimated values.
- ACC aggregates exclude Malta.
- EEA does not include Liechtenstein.

Time series

Chapter 2 and 3 present data for the period 1993-2002. However, data in NewCronos are available from 1981 onwards, but differences exist according to the variables and the institutional sectors. Not all years are completed, and therefore the latest year available for each country is presented in the analysis.

Sources

- Eurostat, *NewCronos*, Theme 9, Domain *RD_EX_P* and Domain *RD_CEC*
- Japan and the United States: OECD, *Main Science and Technology indicators* – MSTI 2002/2.

Methodology in the Candidate Countries

Most Candidate Countries have introduced Frascati methodology from 1994-98.

More detailed information regarding the specific developments of each country are available in Eurostat's Reference database *NewCronos*.

For these and further information on definitions and explanatory notes, see:

- Metadata in Eurostat's reference database *NewCronos*, Theme 9, Domain *RD_EX_P* and Domain *RD_CEC* or in
- ***Statistics on Science and Technology: annual statistics – Data 1991-2001.***

Human Resources in Science and Technology — HRST

Data on Human Resources in Science and Technology — HRST — can improve our understanding of both the demand for, and supply of, highly qualified personnel. The data in this publication focuses on two main aspects: stocks and flows. The former serves to show the needs and the current situation of the labour force and the latter indicates to what degree this demand is likely to be met in the future by looking at the current participation and graduation output of the educational systems.

The general recommendations for the collection of HRST data are laid down in the ***Canberra Manual*** (4), where HRST is defined as a person fulfilling one of the following conditions:

- successfully completed education at the third level in a S&T field of study (ISCED '97 version levels 5a, 5b or 6) or;
- are not formally qualified as above but are employed in a S&T occupation where the above qualifications are normally required (ISCO '88 COM codes 2 or 3).

The conditions of the above educational or occupational requirements are considered according to internationally harmonised standards:

- the ***International Standard Classification of Education*** — ISCED;
- the ***International Standard Classification of Occupation*** — ISCO (5).

Stocks

Stocks provide information on the number of HRST at a particular point in time. In this publication, stock data relate to the employment status as well as the occupational and educational profiles of individuals in quarter 2 of any given year.

HRST stock data and their derived indicators are extracted and built up using data from the EU Labour Force Survey. The EU Labour Force Survey, like all surveys, is based upon a sample of this population. Therefore, the results are subject to the usual types of errors associated with sampling techniques as well as a number of other non-sampling errors, for example, non-response, miscoding, etc. All results conform to Eurostat guidelines on sample-size limitations and are therefore not published if the degree of sampling error is likely to be high.

(4) *Manual on the Measurement of Human Resources devoted to S&T* — ***Canberra Manual***, OECD, Paris, 1994.

(5) Education data follow the ***International Standard Classification for Education*** — ISCED, whilst occupation data follow the ***International Standard Classification for Occupation*** — ISCO.

The basic categories of HRST are as follows:

Category	*People that have/are*
• **HRST**: Human Resources in Science and Technology	• successfully completed education at the third level in an S&T field of study (6) – ISCED '97 version levels 5a, 5b or 6 – **or** • are not formally qualified as above but are employed in an S&T occupation where the above qualifications are normally required (ISCO '88 COM codes 2 or 3).
Sub-categories of HRST	*People belonging to HRST that have/are*
• **HRSTO**: Human Resources in Science and Technology – Occupation	• employed in an S&T occupation (ISCO '88 COM codes 2 or 3).
• **HRSTE**: Human Resources in Science and Technology – Education	• successfully completed education at the third level in an S&T field of study (6) – ISCED '97 version levels 5a, 5b or 6.
• **HRSTC**: Human Resources in Science and Technology – Core	• successfully completed education at the third level in an S&T field of study (6) – ISCED '97 version levels 5a, 5b or 6 – **and** are employed in a S&T occupation (ISCO '88 COM codes 2 or 3).
• **S&E**: Scientists and Engineers	• physical, mathematical and engineering occupations (ISCO '88 COM code 21); • life science and health occupations (ISCO '88 COM code 22).
• **HRSTU**: Human Resources in Science and Technology – Unemployed	• successfully completed education at the third level in an S&T field of study (6) – ISCED '97 version levels 5a, 5b or 6 – **and** are unemployed.
• **NHRSTU**: Unemployed non-HRST	• unemployed and without successfully completed education at the third level in a S&T field of study (6) (ISCED '97 version levels 5a, 5b or 6).

Inflows

HRST inflows are the number of people who do not fulfil any of the conditions for inclusion in HRST at the beginning of a time period but gain at least one of them during the period.

The number of graduates from a country's higher education system represents the main inflow into the national stock of HRST.

HRST education inflow data are extracted from the Eurostat Education database and are collected via the Unesco/OECD/Eurostat questionnaire on education, the conditions of which are considered according to the *International Standard Classification of Education* – ISCED.

(6) Note that according to the ***Canberra Manual***, the seven broad S&T fields of study are 'Natural sciences', 'Engineering and technology', 'Medical sciences', 'Agricultural sciences', 'Social sciences', 'Humanities' and 'Other fields', ***Canberra Manual***, § 71.

The International Standard Classification of Education — ISCED 97

Levels of tertiary education

• **ISCED level 5A**	• programmes that are largely theoretically based and are intended to provide sufficient qualifications for gaining entry into advanced research programmes and professions with high skill requirements.
• **ISCED level 5B**	• programmes that are generally more practical/technical/occupationally specific than ISCED 5A programmes.
• **ISCED level 6**	• this level is reserved for tertiary programmes that lead to the award of an advanced research qualification. The programmes are devoted to advanced study and original research.

S&E (field of study)

Title	*Short name*	*Description*	*ISCED subject codes*
• **Science and Engineering**	• S&E	Life sciences, Physical sciences, Mathematics and statistics, Computing, Engineering and engineering trades, Manufacturing and processing, Architecture and building.	42, 44, 46, 48, 52, 54, 58

The International Standard Classification of Occupations — ISCO (S&T occupations)

• **ISCO 2** (professionals)	• occupations whose main tasks require a high level of professional knowledge and experience in the fields of physical and life sciences, or social sciences and humanities.
• **ISCO 3** (technicians and associate professionals)	• occupations whose main tasks require technical knowledge and experience in one or more fields of physical and life sciences, or social sciences and humanities.

The user should note that definition of S&T occupations constitutes a certain deviation from the recommendations laid down in the Canberra Manual. In addition to ISCO major groups 2 and 3 the Canberra Manual proposes to also consider as HRST: production and operations managers, other specialist managers, managers of small enterprises (ISCO 122, 123 and 131) that may work in the field of S&T but are not included in the term HRST as used here (but they are included in HRST(E) if they have successfully completed third level education). The limitation applied here is however justified as a pilot survey conducted in 1995 tested the validity of the original definitions for HRST and the results indicated that, for the EU, including these certain managerial occupations distorted the results significantly, due to variations between countries in the treatment and classification of managers.

Non-national students

As a foreign student is defined as someone not having the citizenship of the country in which he/she is educated, overestimation of non-national students may exist in some countries where permanently resident second generation migrants with foreign nationalities constitute an important group of students.

Breakdown by sector of activity

HRST data by sector of activity are collected according to the *Statistical classification of economic activities in the European Community* — NACE Rev. 1.1.

According to their global technological intensity, the OECD and Eurostat have agreed on a classification of manufacturing sectors. Although the agreed classification defines each sector using up until the third digit level of the NACE, the *EU Labour Force Survey* only allows reporting of NACE at the 2 digit level. Following a similar logic as for manufacturing, Eurostat proposes the classification of services sectors in various categories by looking at their knowledge intensity and within that their high technology usage and their market orientation.

Therefore, the sector groups used in Chapter 4 are defined as follows:

Description	*NACE Rev 1.1 codes (1)*
• Agriculture, hunting, forestry, fishing, mining and quarrying	• 01 to 14
• Utilities and construction	• 40, 41 and 45
• Low technology	• 15 to 22 and 37
• Medium-low technology	• 23, 25 to 28 and 36
• Medium-high tech manufacturing	• 24, 29, 31, 34 and 35
• High tech manufacturing	• 30, 32 and 33
• Knowledge-intensive high technology services	• 64, 72 and 73
• Knowledge-intensive market services (excl. financial intermediation and high tech services)	• 61, 62, 70, 71 and 74
• Knowledge-intensive financial services	• 65, 66 and 67
• Other knowledge-intensive services	• 80, 85 and 92
• Less-knowledge-intensive market services	• 50, 51, 52, 55, 60 and 63
• Other less-knowledge-intensive services	• 75, 90, 91, 93, 95 and 99.

(1) See definitions of each code in ***Statistical classification of economic activities in the European Community*** — NACE Rev. 1.1.

Time series

Data are available in many countries from 1994 onwards, but differences exist and certain years are missing. Users should note that the existence of data in this *NewCronos* domain further depends on their reliability.

Sources

- Eurostat, *NewCronos*, Theme 9, Domain **HRST**,
- *Third European Report on Science and Technology Indicators*, Directorate-General for Research, 2003.

For further data and additional methodological notes please refer to:

- Eurostat's reference database ***NewCronos***, Theme 9, Domain **HRST**.

Patents

Patents reflect part of a countries' inventive activity. Patents also show the country's capacity to exploit knowledge and translate it into potential economic gains. In this context, indicators based on patent statistics are widely used to assess the inventive performance of the country or regions.

The grounds for the assumption that a patent represents a codification of inventive activity rely on the novelty, utility and inventiveness that an invention requires to be subject to be patented. On the basis of this assumption, Eurostat collects patent statistics to build up indicators of R&D output.

Eurostat's patents database contains data on patent applications to the European Patent Office — EPO — and patents granted by the United States Patent and Trademark Office — USPTO. In addition Chapter 5 looks at data on triadic patent families, which originate from the ***OECD's MSTI database***. Due to methodological differences in the manner of processing the data, no cross sectional comparisons are advisable between the EPO, USPTO and patent family data. Methodological issues specific to each type of data are explained below:

Patent applications to the EPO by year of filing

Data in Eurostat's EPO database refers to patent applications to the EPO by year of filing, which include both applications filed directly under the European Patent Convention and applications filed under the ***Patent Co-operation Treaty*** and designating the EPO — Euro-PCT — for protection. The regional distribution of patent applications is assigned according to the inventor's place of residence. If one application has more than one inventor, the application is divided equally among all of them and subsequently among their regions, thus avoiding double counting.

Data in this collection are given broken down according to the ***International Patent Classification*** — IPC, which assigns an invention to an IPC-class according to its function or intrinsic nature or its field of application. If a patent is assigned to more than one IPC code, the application is equally divided among all the IPC-sub-classes in order to avoid double counting. Regional data are given according to the ***Nomenclature of Territorial Units for Statistics — NUTS*** classification, 1998 version.

The EPO collection contains data not only for total patent applications but also for applications in the high technology fields. The definition of high tech followed by Eurostat is that of the ***Trilateral Statistical Report***, a joint publication of the EPO, the JPO and the USPTO (1999). Here, six technical fields are defined as high technology and are constructed by aggregating the following IPC codes:

1.	Computer and automated business equipment:	B41J+G06+G11C,
2.	Micro-organism and genetic engineering:	C12M+C12N+C12P+C12Q,
3.	Aviation:	B64,
4.	Communication technology:	H04,
5.	Semi-conductors:	H01L,
6.	Lasers:	H01S.

EPO data are available from 1989 to 2001, 2001 data being provisional.

For further information on definitions and explanatory notes concerning EPO patent data see:

- Metadata in Eurostat's reference database ***NewCronos***, Theme 9, Domain **PATENTS**, Collection PAT_EU and
- ***Statistics on Science and Technology: annual statistics — Data 1991-2001.***

Patents granted by the USPTO by year of publication

Data on patents granted by the USPTO refer to patents granted, and not to applications as is the case for data coming from the EPO. Also, the reference year corresponds to the year of publication as opposed to the year of filing used for EPO data. In this context, data in these two collections are not comparable. USPTO data are available from 1991 to 2001.

For further information on definitions and explanatory notes concerning USPTO patent data see:

- Metadata in Eurostat's reference database *NewCronos*, Theme 9, Domain **PATENTS**, Collection **PAT_US** and
- *Statistics on Science and Technology: annual statistics — Data 1991-2001*.

Triadic Patent families by priority year

Triadic Patent families data was obtained from the *OECD's MSTI database*. The patent families presented in this publication refer to *triadic* families: i.e. a patent is a member of the patent families if and only if it has been applied for and filed at the *European Patent Office* — EPO — and the *Japanese Patent Office* — JPO — and if it has been granted by the *US Patent & Trademark Office* — USPTO. Patent families, as opposed to patents, are provided with the intention of improving international comparability (the *home advantage* is suppressed, the values of the patents are more homogeneous).

It has to be noted that data on triadic patent families is presented by priority year, i.e. the year of the first international filing of a patent. This increases the drawback of traditional patent counts with respect to timeliness and therefore latest available data refers to 1998 only.

For further methodological notes please refer to:

- Compendium of Patent Statistics, OECD, 2003.

Employment in high technology in Europe

Sources

Employment data presented in Chapter 6 originate from Eurostat's *Employment in High Tech database* – EHT. Eurostat's EHT database includes data on employment in high technology and medium-high technology manufacturing sectors, knowledge-intensive service – *KIS* – sectors, high technology service sectors, other KIS sub-sectors and reference sectors. Employment in high tech data and derived indicators are extracted and built up using data from the *EU Labour Force Survey*.

The database covers a time series from 1994 onwards, but differences exist and certain years are missing. Existence of data further depends on their reliability. Data are currently available at the national and regional levels – NUTS '99 levels 1 and 2 – for the 15 Member States of the European Union. Data at the national level are also available for some Acceding and Candidate Countries, Iceland, Norway and Switzerland.

The data presented are based on the *Statistical classification of economic activities in the European Community*, NACE Rev.1.1, 1996 both at the national and regional levels. Regional data are presented according to the *Nomenclature of Territorial Units for Statistics*, 1998, developed by Eurostat at the NUTS levels 1 and 2.

Classification of high tech and knowledge-intensive sectors

As for chapter 4 on HRST, high technology sectors in manufacturing are defined according to their global technological intensity, following the classification agreed between the OECD and Eurostat. Similarly, Eurostat's proposed classification of services sectors is also followed in the employment in high tech database. As the employment in high tech data presented in Chapter 6, Section 6.2. is also based on the *EU Labour Force Survey* data, the group of sectors are defined also at the 2 digit level of the NACE.

The groups of sectors presented in this chapter are defined below:

Description	*NACE Rev 1.1 codes (1)*
• Total	• All sectors
• Manufacturing	• Section D
• High tech and medium-high tech manufacturing	• 30, 32, 33, 24, 29, 31, 34 and 35
• High tech manufacturing	• 30, 32 and 33
• Services	• Sections G to Q
• Knowledge-intensive services	• 61, 62, 64, 65, 66, 67, 70, 71, 72, 73, 74, 80, 85 and 92
• Knowledge-intensive high technology services	• 64, 72 and 73.

(1) See definitions of each code in ***Statistical classification of economic activities in the European Community*** — NACE Rev. 1.1.

Due to the lack of employment data at the 2-digit level, employment in high tech and knowledge-intensive sector indicators for MT, PL and TR may not be calculated and therefore are not presented in this publication.

Quality of the data

The guidelines on the sample size reliability of the data established by the *EU Labour Force Survey* are applied to the EHT database and therefore regions for which reliability levels do not permit publication appear as not available and are flagged as unreliable. Regions for which reliability levels define the data as unreliable but allow for publication are included in the rankings and flagged as unreliable.

For further methodological notes please refer to:

- Eurostat's reference database ***NewCronos***, Theme 9,
  Domain *Employment in high and medium-high technology sectors — EHT*.

Value added and Labour productivity

Source

Data on value added and labour productivity in Chapter 6 were obtained from Eurostat's *Structural Business Statistics — SBS* — database.

Value added at factor cost

Value added at factor cost is the gross income from operating activities after adjusting for operating subsidies and indirect taxes.

Labour productivity

Labour productivity refers to the gross value added per person employed.

Classification of high tech and knowledge-intensive sectors

SBS data are available at the three-digit level and therefore indicators based on this source follow the strict definitions of high tech and medium-high tech manufacturing sectors agreed by the OECD and Eurostat. In this sense, high tech manufacturing indicators on Table 6.12. include Classes 24.4 — Manufacture of pharmaceuticals, medicinal chemicals and botanical products — and 35.3 — Manufacture of aircraft and spacecraft, whereas high tech and medium-high tech manufacturing indicators exclude Class 35.1 — Building and repairing of ships and boats.

High tech trade

High tech products

In order to analyse the competitive and trade performance of high tech trade markets, two main approaches (by sector and product) are used to identify the technology-intensive industries and products.

In the sectoral approach, the OECD currently identifies four high tech industry groupings based on R&D intensity:

- Aerospace,
- Computers and Office Machinery,
- Electronics-telecommunications, and
- Pharmaceuticals.

However it should be noted that not all high tech industry products are high tech products – some of them may be medium or low tech products.

A product approach was recently devised to complement the sectoral approach. It opens the way to far more detailed analysis of trade and competitiveness. The product list is based on the calculations of R&D intensity by groups of products (R&D expenditure/total sales covering six countries). The products classified as high technology products are listed on table below. The exports and imports of these products comprise high tech trade.

For the purposes of this chapter, the product approach is used to analyse the evolution of high tech trade which makes up a considerable proportion of total trade in many advanced economies.

High technology products are defined as listed below:

List of high technology Products	*SITC Rev.3*
Aerospace	7921+7922+7923+7924+7925+79293+(714-71489-71499)+87411
Computers-office machines	75113+75131+75132+75134+(752-7529)+75997
Electronics-telecommunications	76381+76383+(764-76493-76499)+7722+77261+77318+77625+7763+7764+7768+89879
Pharmacy	5413+5415+5416+5421+5422
Scientific Instruments	774+8711+8713+8714+8719+87211+(874-87411-8742)+88111+88121+88411+88419+89961+89963+89967
Electrical machinery	77862+77863+77864+77865+7787+77844
Chemistry	52222+52223+52229+52269+525+57433+591
Non-electrical machinery	71489+71499+71871+71877+72847+7311+73135+73144+73151+73153+73161+73165+73312+73314+73316+73733+73735
Armament	891

Source: OECD, STI working paper 1997/2.

Units

Imports and exports are expressed in current Euro. In the absence of an appropriate trade deflator it was decided to use trade data at current prices. Figures reported for the total European Union exclude the intra-EU trade. Nevertheless, for the individual EU countries intra-EU trade is included.

Sources

All high tech trade data relating to the EU countries are based on data extracted from the *COMEXT* database – Eurostat's database of official statistics on EU's external trade and trade between EU Member States. This database includes imports and exports data flows with Member States and third countries as reported by the EU countries only. Trade data reported by third countries, including the Acceding Countries – ACC, were extracted from the UN statistical office's *Comtrade* database. The information from the latter is used to calculate the world total trade flows and high tech trade as well as high tech market shares. It should be noted that in calculating the world total trade flows, the trade data reported by the EU countries in the *COMEXT* database were added to the trade data reported by third countries in the *Comtrade* database to obtain the world total trade.

Time series

Trade data in Chapter 6 cover the reference period 1996-2001.

World totals

World totals used in this chapter in order to calculate world market shares are estimated. The world totals were calculated as the sum of the following list of 87 countries

- Belgium
- Denmark
- Germany
- Greece
- Spain
- France
- Ireland
- Italy
- Luxembourg
- the Netherlands
- Austria
- Portugal
- Finland
- Sweden
- United Kingdom
- Czech Republic
- Estonia
- Cyprus
- Lithuania
- Hungary
- Malta
- Poland
- Slovenia
- Slovak Republic
- Romania
- Iceland
- Norway
- Algeria
- Argentina
- Australia
- Bangladesh
- Barbados
- Bolivia
- Brazil
- Canada
- Chile
- China
- Colombia
- Costa Rica
- Croatia
- Ecuador
- Greenland
- Grenada
- Guadeloupe
- Guatemala
- Honduras
- Hong Kong
- India
- Indonesia
- Israel
- Jamaica
- Japan
- Jordan
- Kenya
- Kuwait
- Macao
- Madagascar
- Malawi
- Martinique
- Mauritius
- Maylasia
- Mexico
- Morocco
- New Zealand
- Nicaragua
- Oman
- Pakistan
- Paraguay
- Peru
- the Philippines
- Reunion
- Saudi Arabia
- Singapore
- St Kitts and Nevis
- St Lucia
- South Africa
- South Korea
- Sri Lanka
- Switzerland
- Taiwan
- Thailand
- Trinidad tbg
- Tunisia
- Turkey
- United States
- Uruguay
- Venezuela
- Zimbabwe.

Abbreviations and symbols

Statistical symbols and abbreviations

-	not applicable or real zero or zero by default
0	less than half of the unit used
:	not available
..	confidential data
%	percentage
p	provisional value
e	estimated value
s	Eurostat estimate
r	revised value
f	forecast
b	break in series
u	unreliable
:u	extremely unreliable data
fax	facsimile number
No	number
p.	page
Tel	telephone number
1990-92	period of several calendar years (e.g. from 1.1.1990 to 31.12.92)

Patents — High tech group titles

AVI	Aviation
CAB	Computer and automated business equipment
CTE	Communication technology
LSR	Lasers
MGE	Micro-organism and genetic engineering
SMC	Semi-conductors

Abbreviations

- **A**

AAGR .... annual average growth rate
ACC .... Acceding Countries
AGR .... annual growth rate

- **B**

BERD .... expenditure on R&D in the Business enterprise sector
BES .... Business enterprise sector
BMBF .... Bundesministerium für Bildung, Wissenschaft, Forschung und Technologie – DE

- **C**

CBS .... Statistics Netherlands
CBSTII .... Common Basis for Science, Technology and Innovation Indicators
CD-ROM .... compact disc read-only memory
CEC .... Commission of the European Communities
CEPS/INSTEAD .... Centre d'Études de Populations, de Pauvreté et de Politiques Socio-Économiques/ International Networks for Studies in Technology Environment, Alternatives, Development – LU
CERIS .... Instituto di ricerca sull'empresa e lo sviluppo – IT
CNR .... Consiglio Nazionale delle Ricerche – IT

- **D**

DG .... directorate-general
DG RTD .... Research Directorate-General
DTI .... The Department of Trade and Industry

- **E**

EC .... European Community/Communities
EEA .... European Economic Area
EHT .... Employment in high tech
EPO .... European Patent Office
ESA .... European system of integrated accounts
EU LFS .... European Union Labour Force Survey
EU/EU-15 .... European Union
EUR .... Euro
EUR-12 .... Eurozone – BE, DE, EL, ES, FR, IE, IT, LU, NL, AT, PT, FI
Eurostat .... Statistical Office of the European Communities

• F

FhG-ISI .... Fraunhofer Institut für Systemtechnik und Innovationsforschung
FP .... Framework programme
FTE .... full-time equivalent

• G

GBAORD .... Government budget appropriations or outlays on R&D
GDP .... gross domestic product
GERD .... gross domestic expenditure on R&D
GISCO .... geographic information system for the Commission — Eurostat
GOV .... government sector
GUF .... General University Funds

• H

HC .... head count
HES .... Higher education sector
HRST .... Human resources in science and technology
HRSTC .... Human resources in science and technology — Core
HRSTE .... Human resources in science and technology — Education
HRSTO .... Human resources in science and technology — Occupation
HRSTU .... Human resources in science and technology — Unemployed

• I

INE .... Instituto Nacional de Estadística — ES
IPC .... International patent classification
ISBN .... international standard book number
ISCED .... International standard classification for education
ISCO .... International standard classification of occupations
ISTAT .... Istituto Nazionale di Statistica — IT
IT .... information technology

• J

JPO .... Japanese Patent Office

• K

KIS .... Knowledge-intensive services

• L

LF .... labour force
LFS .... Labour Force Survey

M

MENRT .... Ministère de l'éducation nationale, de la recherche et de la technologie – FR

Mio .... million

Mio EUR .... millions of euro

MSTI .... Main Science and Technological Indicators – OECD

N

NABS .... Nomenclature for the analysis and comparison of science budgets and programmes

NACE .... General industrial classification of economic activities within the European Communities

NESTI .... National Experts on Science and Technology Indicators

NewCronos .... Eurostat's statistical reference database

NHRSTU .... unemployed non-HRST

NIFU .... Norwegian Institute for Studies in Research and Higher Education

NUTS .... Nomenclature of territorial units for statistics

O

OCT .... Observatório das Ciências e das Tecnologia – PT

OECD .... Organisation for Economic Cooperation and Development

ONS .... Office for National Statistics – UK

OPOCE .... Office for Official Publications of the European Communities

OST .... Observatoire des Sciences et des Techniques – FR

P

PCT .... Patent Cooperation Treaty

PhD .... Philosophiae Doctor – doctor of philosophy

PNP .... private non-profit sector

PPP .... purchasing power parities

PPS .... purchasing power standard

R

R&D .... research and development

REIST-3 .... Third European Report on Science and Technology Indicators

RSE .... researchers

RSI .... Relative specialisation index

RTD .... research and technological development

Abbreviations and symbols

• S

SBS .... Structural Business Statistics

SITC .... Standard international trade classification

STATEC .... Service Central de la Statistique et des Études Économiques — LU

STI .... Science, technology and innovation

S&E .... Scientists and engineers

S&T .... Science and technology

• U

UN .... United Nations

UOE .... Unesco/OECD/Eurostat

USPTO .... United States Patent and Trademark Office

• W

WIPO .... World Intellectual Property Organisation

• Z

ZEW .... Zentrum für Europäische Wirtschaftsforschung — GmbH — DE

Countries

• EU-15

BE .... Belgium

DK .... Denmark

DE .... Germany

EL .... Greece

ES .... Spain

FR .... France

IE .... Ireland

IT .... Italy

LU .... Luxembourg

NL .... the Netherlands

AT .... Austria

PT .... Portugal

FI .... Finland

SE .... Sweden

UK .... the United Kingdom

- **Acceding Countries — ACC**

CZ	Czech Republic
EE	Estonia
CY	Cyprus
LV	Latvia
LT	Lithuania
HU	Hungary
MT	Malta
PL	Poland
SI	Slovenia
SK	Slovak Republic

- **Other Candidate Countries**

BG	Bulgaria
RO	Romania
TR	Turkey

- **Other countries**

CA	Canada
CH	Switzerland
CN	China
ID	Indonesia
IS	Iceland
HK	Hong Kong
JP	Japan
KR	South Korea
LI	Liechtenstein
MX	Mexico
MY	Malaysia
NO	Norway
PH	Philippines
SG	Singapore
TH	Thailand
TW	Taiwan
US	United States

Abbreviations and symbols